Understanding Workers' Compensation

A Guide for Safety and Health Professionals

Kenneth M. Wolff, D.C.

Government Institutes
An imprint of
The Scarecrow Press, Inc.
Lanham, Maryland • Toronto • Oxford

Published in the United States of America
by Government Institutes, an imprint of The Scarecrow Press, Inc.
A wholly owned subsidiary of
The Rowman & Littlefield Publishing Group, Inc.
4501 Forbes Boulevard, Suite 200
Lanham, Maryland 20706
govinst.scarecrowpress.com

PO Box 317, Oxford, OX2 9RU, UK

Printed in the United States of America.

Library of Congress Cataloging-in-Publication Data

Wolff, Kenneth.
Understanding workers' compensation : a guide for safety and health professionals / Kenneth Wolff.
p. cm.
Includes bibliographical references and index.
ISBN 0-86587-464-6
1. Workers' compensation--Law and legislation--United States.
I. Title.
KF3615.W65 1995
344.73'021--dc20
[347.30421] 95-8770
CIP

Dedication

This book is dedicated to my father, whom I loved and respected, and who always wanted to have a book published and never did.

Table of Contents

Foreword

Although workers' compensation management should be a tool to help reduce injuries and costs, most business executives do not give workers' compensation much thought. Each state requires nearly every private organization to carry workers' compensation insurance, yet American business takes workers' comp for granted. The insurance is seen as a cost of doing business, with little understanding of how the system functions or how costs can be controlled. Managers know little more about it than that premiums come due with regularity, and that the costs associated with it tend to go up each year.

Some, however, correctly feel that they should know more about this subject. Many health and safety professionals now have workers' comp management as part of their responsibilities. Although workers' comp is now part of the safety curriculum at many colleges and universities, the majority of health and safety professionals working today have little or no knowledge of the subject. Short of reading hundreds of pages of fine print of each state's workers' comp regulations, it has not been possible to easily grasp what workers' comp is, how it works, and how to manage it. Dr. Kenneth Wolff has now provided the means to do so.

The author of this book is a doctor of chiropractic medicine. What can he possibly tell us about the book's subject, you may ask. Aren't all books on workers' comp usually written by lawyers and CPCU's? Millions of words have been written on the subject, but in my many years in insurance and industry, I have not encountered a better book on this subject. Dr. Wolff does what other authors have failed to do, namely, presents a great deal of practical information, in plain English instead of legalese, which can be used to understand the system (including how regulations vary from state to state and how they are the same).

Understanding Workers' Compensation explains the history of workers' compensation and describes what types of injuries workers' comp covers, how eligibility is determined, and what companies' responsibilities are. But the practical insights Dr. Wolff provides are what make this book unique—clearly distinct from other books which primarily cover the legal aspects of workers' compensation. This book should be a primary source of help to anyone who deals with workers' compensation.

Dr. Wolff offers practical advice on controlling workers' compensation costs. His chapter on ergonomics, for example, provides sound advice on eliminating many of the injuries that plague American business. Tips on getting injured employees back to work are extremely valuable, as is his guidance on the basic process for handling injured workers. And his comparison of the costs involved in different ways of treating injuries is thought-provoking.

In short, Dr. Wolff shows us the forest, helps us examine some of the more important trees, and then leads us through to the other side. Progressive managers and safety professionals would do well to follow the path he has prepared.

Keith E. Barenklau, PhD, PE, CSP
Program Director and Professor
Safety Technology Program
Marshall University

Acknowledgments

It is nearly impossible to list all the people who have helped me in many ways, in the preparation of this text. So if I have not indicated your name here, please realize that you have my deepest thanks and appreciation.

Kathy Bazan, for the idea, help, and the belief that I could actually do this. Alex Padro, for taking the gamble; Dr. Craig Cook, for both moral support and feedback; Mariah, my daughter, whose joyous smile brings sunshine to my life; and last and definitely not least, my wife Kathy, for her support, understanding, feedback, and love as I spent many days and hours at the computer putting this book together.

Chapter 1

Where Did It All Begin?

Workers' compensation is a realm of business that every employer and employee should know about. It is probably the most underrated and misunderstood area of business, yet other than taxes, it is probably one of the most costly.

In the United States, all workers are covered by state workers' compensation laws, except those that are covered by federal programs. The Longshore and Harborworkers' Compensation Act, applicable to maritime workers on navigable waters of the United States, and the Federal Employees' Compensation Act, applicable to civilian workers of the United States government, provide coverage on a federal level. Workers' Compensation is a system or a group of systems in many different jurisdictions of the United States and other countries that allows cash benefits, rehabilitation and medical costs to be paid to workers injured or killed on the job, or to those who have been disabled from an on-the-job injury or illness.

As an employer, it is imperative to know how the system works, what part you play in the system, what to do when things go wrong, and how to keep the system—from the employer's end—running as smoothly and efficiently as possible. This text presents both success and horror stories found in workers' compensation.

The system began in the late 19th and early 20th centuries, around the time of the Industrial Revolution, when employees grew weary of being abused and taken advantage of by their employers. In those days, employees received no compensation whatsoever should they be injured or sick, or die. These employees had to work under any circumstances; otherwise, they wouldn't get paid and they could lose their jobs. Their only recourse against their employers was tort law, and generally it was proven that the injury was not the fault of the employer; thus the employee was out of luck. It was nearly impossible for the injured worker to prove that the injury was caused by the employer or the job site.

In those days, the employers had total and complete control over their employees. They forced the workers to work under bad conditions, paid them poorly, and made them work whether they were sick or lame, injured or near death, because they were the bosses and the employees were practically slaves—poorly paid ones at that.

Workers had no rights. They had no benefits, no ability to recoup losses if injured, no rehabilitation, no after-injury job training. These features were brought about by the advent of workers' compensation.

Workers' compensation laws became necessary as a result of two developments in the workplace: a dramatic increase in work-related accidents due to the rise of the factory system, and a decline in common-law remedies for the employees' injuries.[1]

"A correctly balanced underlying concept of the nature of workers' compensation is indispensable to an understanding of current cases and to a proper drafting and interpretation of compensation acts. Almost every major error that can be observed in the development of compensation law, whether judicial or legislative, can be traced either to the importation of tort ideas, or, less frequently, to the assumption that the right to compensation resembles the right to the proceeds of a personal insurance policy."[2]

"In the early 1900s, Germany did a study of employer's liability in workers' injuries. At that time, the employer used any means possible to not have to pay any benefits to the injured worker. Figure 1.1 below shows the results of this study, by way of classification of causes of accidents.

Classification of Causes of Accidents

Cause	Percentage
1. Negligence or fault of employer	16.81%
2. Joint negligence of employer and injured worker	4.66%
3. Negligence of fellow worker	5.28%
4. Acts of God	2.31%
5. Fault or negligence of injured worker	28.89%
6. Inevitable accidents connected with employment	42.05%

Figure 1.1

Source: Arthur Larson, *The Law of Workmen's Compensation.* (New York: Matthew Bender & Co., Inc., 1994).

"It is at once apparent that, with numbers 2 and 3 barred by common-law defenses, only under number 1 is there any possibility of employer liability; accordingly, the employee [under] common law was remediless without question in [83.19%] of all cases.

"What of the remaining 16.81%? The defense assumption of risk might still apply, for even where the employer was at fault, many cases held that the employee, by continuing to work in spite of the defects or dangers created by the employer, consented to waive [the] employer's obligation." [3] (See figure 1.2 below.)

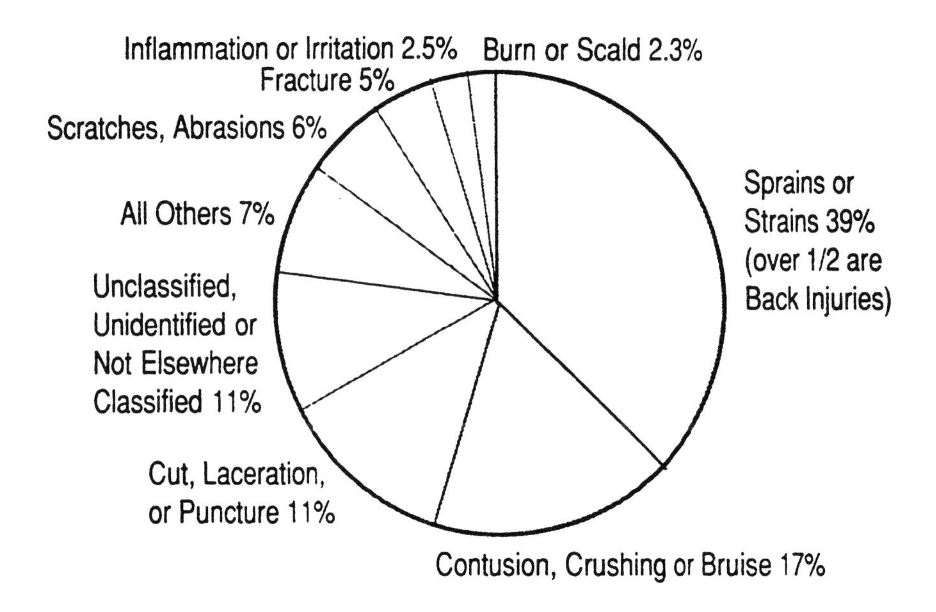

Figure 1.2: Disabling Work Injuries by Nature of Injury

Source: Canadian Centre for Occupational Health and Safety, *Nature of Injury by Occupation.* (Toronto: Canadian Centre for Occupational Health and Safety, 1985).

"When studies were first done in America, prior to compensation legislation, they usually showed what injured workers actually received under the old system. The conclusions were usually shocking. The Illinois commission investigated 5,000 industrial accidents, and found that of 614 death cases, 214 families received nothing, and were engaged in pending litigation in 111 [cases]. The other cases settled for very minor amounts, averaging a few hundred dollars. The report of the New York commission contains a number of similar tabulations, of which the table on fatal industrial accidents in New York City in 1908 is typical: Of 74 cases whose disposition was known, there was no compensation in 43.2%, and compensation under $500 in 40.5%, with only 16.3% receiving between $500 and $5,000. Even these figures do not tell the whole story, since attorneys' fees had to come out of these meager settlements and averaged a fourth to a third of the amount received." [4]

During the early 1900s, states began to set up commissions to investigate workers' compensation. Around 1910, the commissions met and established a Uniform Workmen's Compensation Law.

The employer has protections under the workers' compensation act or law in each particular state. The employer cannot be sued civilly for an injury occurring on the job, assuming that the employer is acting properly within the system.

"The employers' rights to immunity may not apply in all circumstances. Below are a few examples of exemptions to the employers' immunity:

- Injuries not covered by workers' compensation
- Injuries sustained by an employee of a noncomplying employer
- Injuries caused by the employers' intentional act
- Injuries sustained while the employer and employee entered into a separate relationship or "dual capacity" independent of their master/servant relationship
- Discharge, demotion, or other punitive action by an employer taken in retaliation for employees' filing workers' compensation claims or otherwise pursuing workers' compensation rights." [5]

Of all the states in the union, Maryland was the first to enact a workers' compensation act, in 1902. It applied only to cases involving a death and was considered unconstitutional in *Franklin v. United Railways and Electric Co.* (1904) because it deprived the worker of the right to jury trial and it violated the separation of powers doctrine (an insurance commissioner, under the executive branch, was performing judicial functions). In 1910, Maryland enacted a voluntary workers' compensation statute, but this proved to be very ineffective because neither the employers nor the employees appeared to care about the law.

Montana enacted a workers' compensation act in 1909, which was primarily designed for the employees in the coal industry. This legislation was declared unconstitutional in *Cunningham v. Northwestern Improvement Co.* (1911), because it said that the employers were not allowed the same protection as the employees, in that there was the potential for double liability (employers had to contribute to the state compensation fund and additionally were open to suit if an employee or beneficiary so elected).

In 1908, Massachusetts enacted a voluntary statute for workers' compensation. It proved to be totally ineffective due to the voluntary nature of the act.

By 1920, all but eight states had adopted compensation acts, with Hawaii being the last state to do so in 1963.

"Statistically, the number of employees covered by some compensation act grew fairly rapidly. In 1915, 41.2% were covered; [in] 1920, 67.4% [were] covered; in 1930, 75.2%; [and] in 1940, 81.5%. By 1989 that percentage had grown to 87% and leveled off there." [6]

Within different states, the percentage of workers covered varies from 100 percent in Hawaii, Maine and other states, to 72 percent in Louisiana and Texas.

"The National Commission on State Workmen's Compensation Laws, which was created by Congress in the Occupational [Safety and Health] Act (OSHA) in 1970, and which consisted of 15 members appointed by the President, submitted its report in July 1972 after about a year of intensive hearings, studies, staff analysis and commission discussions. The essential elements recommended by the commission were: compulsory coverage in all acts; elimination of all numerical and occupational exemptions in coverage, including domestic and farm labor; full coverage of work-related diseases; full medical and physical rehabilitation services without arbitrary limits; broad extraterritoriality provision; elimination of arbitrary limits on duration or total sum of benefits; and a weekly benefit maximum that rises from an immediate 66 2/3% to an ultimate of 200% of average weekly wage in the state." [7]

The federal government never adopted any sanctions because of the difficulty in enforcement and political acceptance. After this time, the most significant change was the increase in maximum benefits. In 1972, the year of the report, the average benefit limit was $56. By 1982, it had risen to $105, where it leveled off, and in 1991 it was down to $101.

In July 1993, the State of California enacted emergency legislation to regain control of its workers' compensation system. Runaway costs in workers' compensation, from the benefits as well as rehabilitation, medical costs and, of course, fraud, caused the system to nearly go bankrupt, forcing many businesses to go illegally uninsured, pay exorbitant rates or leave the state. As of January 1, 1995, California has an open market approach to costs of workers' compensation insurance. For the first time ever in California, insurance rates will no longer be regulated; thus the costs are expected to come down dramatically for all involved.

The changes enacted should have a dramatic effect on this heavily overburdened system. Understand that California is not the only state in this kind of trouble. It is a problem in all states to some degree.

How is the Typical Workers' Compensation System Defined?

As stated in the constitution of the State of California, a complete system of workers' compensation is one which:

- has adequate provisions for the comfort, health and safety, and general welfare of any and all workers and those dependent upon them for support;
- fully provides relief from the consequences of any injury or death incurred or sustained by workers in the course of their employment, irrespective of the fault of any party;
- fully provides for securing safety in places of employment; and
- fully provides for such medical, surgical, hospital, and other remedial treatment as is requisite to cure and relieve the effects of such injury.

Understand that this is California's definition. While each state will opt for its own distinct verbiage in the letter of the law, the same spirit is expressed here.

The Structure of a Workers' Compensation System

It has many parts, many divisions and many different terms or definitions. Let's start with the parts. You have parts devoted to medical, legal, rehabilitation, and disability aspects, along with parts concerning insurance companies, judges, a labor board, paperwork, raters, and a wide variety of claims. Medically speaking, this portion of the system consists of the treating doctors and examining doctors, from both the defense and the plaintiff sides, as well as independent medical examiners.

In the majority of all states, "treating doctors" includes medical doctors, chiropractors, physical therapists, psychologists, psychiatrists and, in some states, acupuncturists. Each profession has its own scope of practice by which it can participate in the arena of workers' compensation.

Medical doctors, chiropractors, psychiatrists and psychologists are considered to be the primary physicians in the workers' compensation (WC) system. The other modalities listed above are used in most states—sometimes with a referral, other times as primary.

Experience has shown that these doctors can play the game in a number of ways. Some of them work for the defense or insurance company side; others work for the plaintiff or injured worker; while others work for both sides, not showing any favoritism.

It is the polarization of these two sides that can cause many of the problems faced by the employer, the insurance company, or the doctor trying to help the injured worker.

You have doctors from the defense saying that little to nothing is wrong with the worker, while the plaintiff's doctors claim that not only is something wrong with the worker, but he is disabled and/or restrictions are necessary, depending on the particulars of the case.

This puts a great financial burden on the system because medical reports are being written by two to four doctors at $200-$1000 apiece (see figure 1.3 below), each one saying almost the opposite of the other. This then causes the need for lawyers, arbitration, or intervention by a WC judge in order to solve the dispute.

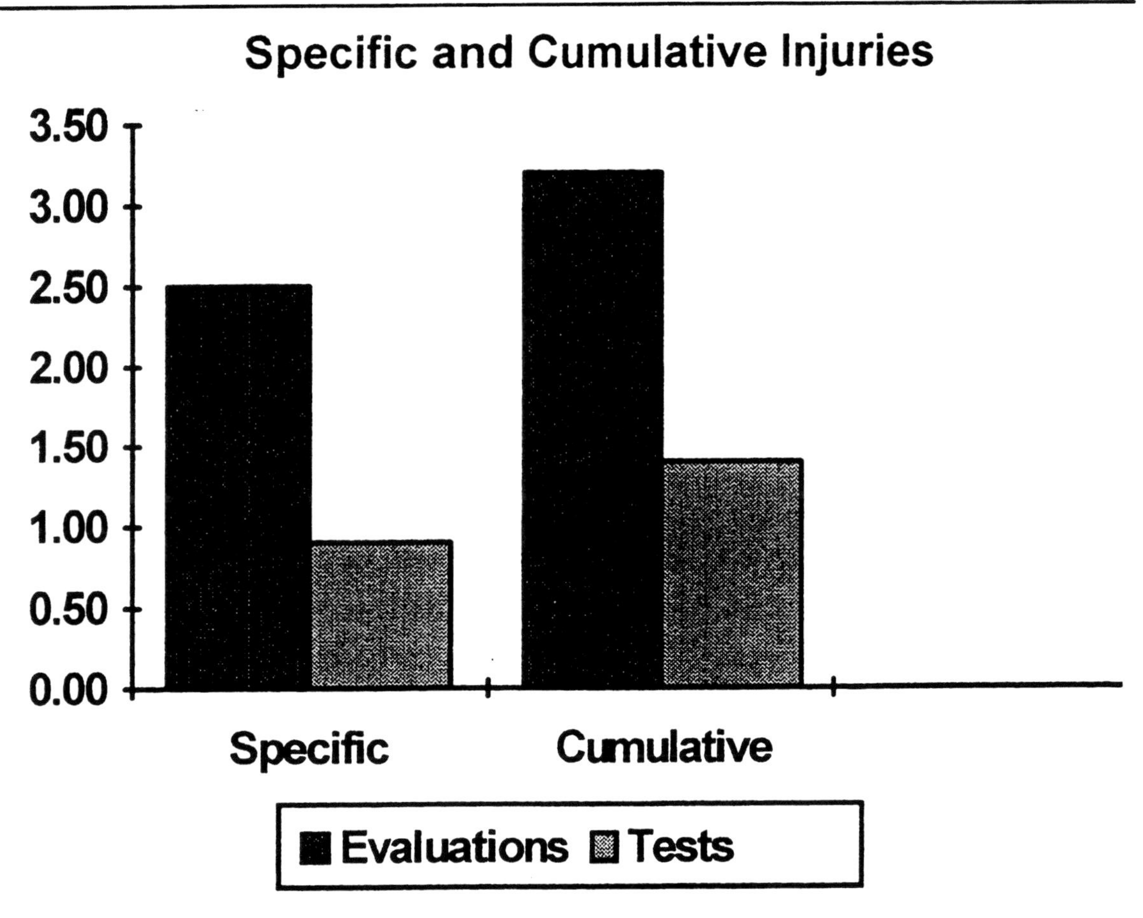

Figure 1.3: Number of Med-Legal Reports Per Claim

Source: M.L. Miller, *Cumulative Injuries and Specific Injuries in California, A Report to Industry.* (San Francisco: California Workers' Compensation Institute, 1995).

And the reality is, the person who has the hardest time during all these disputes is the injured worker. The worker may need medical care, disability payments, etc., which may be put on hold because a doctor paid by the defense may say that nothing is wrong and no benefits are due. At this time, the insurance companies stop paying the worker and/or the worker's doctor. With the added financial burden of a stack of doctor bills, the worker may discontinue treatment, get worse, or become infuriated at his employer for "abandoning him to the wolves."

Doctors within the system should look at the worker as an injured individual. The quality of treatment should not be based on who is paying the doctor. Most injured workers were actually injured on the job (not malingering) and are simply looking for medical relief from the injury. If the worker cannot return to the previous job, the worker seeks rehabilitation or training for a new job and compensation for the lost wages. The worker only wants these things in a reasonable time frame, so that he or she can return to work and become a useful part of society again.

The controversy arises because of the inborn antagonism between the defense and the plaintiff. This medical/legal wrangling is costly to the insurance industry, the employer, the employee and the state.

A typical workers' compensation case in California costs the insurance company more than $25,000 (see figure 1.4 below). This includes medical costs, disability benefits and vocational rehabilitation.

Specific and Cumulative Injuries

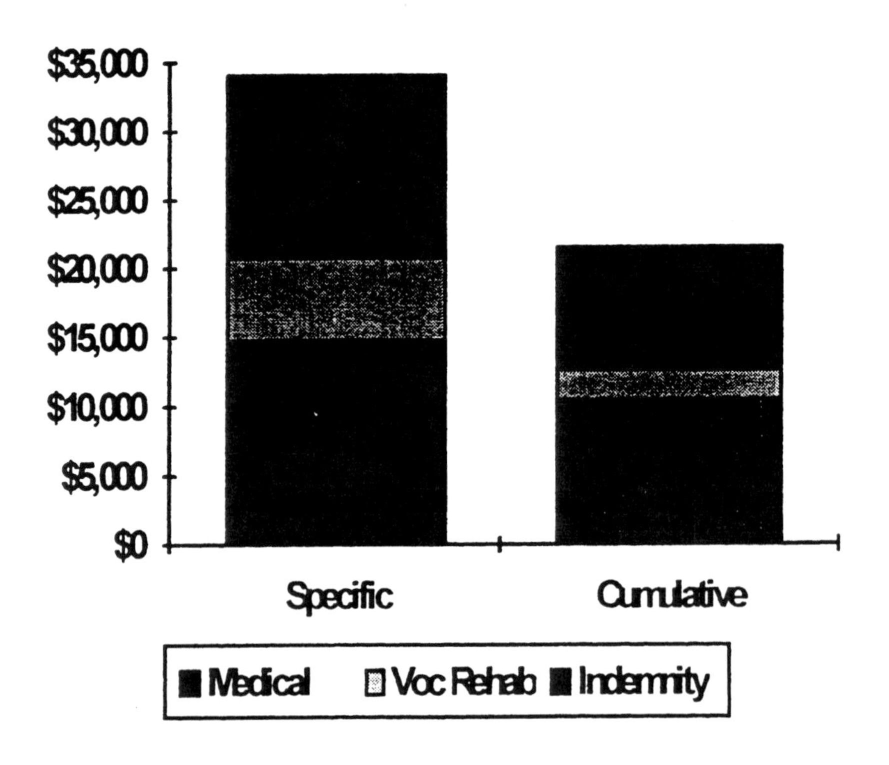

Figure 1.4: Incurred Benefit Costs

Source: M.L. Miller, *Cumulative Injuries and Specific Injuries in California, A Report to Industry.* (San Francisco: California Workers' Compensation Institute, 1995).

The attorney's role in workers' compensation is basically the same as in every avenue of law: try to win. Someone makes a claim, runs into a problem, and then has to have a lawyer to help solve the problem. There are lawyers for both sides. The truth is if the system worked better, there probably would not be as much need for attorneys. The main reasons causing the need for attorneys are that the employer is ignorant of how the system works, is afraid of the system, or simply does not want to comply due to financial considerations.

Judges and the labor board are the final place for any action. This is the place where all final decisions are reached because the parties could not reach a settlement on their own. The judges will use raters to determine the rate of disability for an individual. The administrators are there to handle any additional paperwork and to make administrative decisions regarding labor codes and regulations in each state.

The raters have a very interesting part in all this. In most states, when there is need for a disability rating, the examining or evaluating doctors send in their medical/legal reports on the worker. These written reports usually do not discuss disability ratings in numbers, but are phrased in a verbiage specific to the field. For example, "Mr. Smith has constant slight pain that will increase to moderate with excessive bending, lifting, and stooping. Therefore, it is my recommendation that he be restricted from all lifting that is more than 75 percent of his preinjury lifting ability. The lifting should be limited to no more than 50 pounds, and there should be repetition not to exceed three times per hour."

A rater would look at that language and give the employee a disability rating of 10-20 percent. This data is not of critical benefit to the employer other than giving a basic knowledge of what goes on behind the scenes of a workers' compensation case. It also helps the employer when dealing with work restrictions that may have been set up in the report.

Paperwork, which in some states can be excessive, is just a routine part of any system; workers' compensation is no exception. There are forms for letting the insurance company know there was an injury. There are forms to tell the employer if the employee has chosen a predesignated physician, forms to announce when the worker is back to work, off work, limited in duty, returned to regular duty, and many more.

Filing a claim sets everything in motion. As stated by Jeffrey Nackley, "A workers' compensation claim is an application by an individual, under a state or federal program, for compensation and other benefits for a medical condition that resulted from work. Most states differentiate among injury or accident claims, occupational disease claims, and death claims." [8]

There are five basic types of claims.

- The two most common types are for injury and/or accident, often arising from sudden or traumatic events.
- Occupational diseases are often specific to a particular trade or industrial process. Examples would be black lung disease for coal miners but not secretaries, and inhalation of toxic pesticides by farm workers but not too many college professors, as well as a number of environmental diseases associated with the workplace.
- Death claims are very obvious. These claims provide benefits to the surviving spouse and children.
- Lastly, there are safety code violations. These kinds of claims can give the employee an additional benefit if the injury or illness was caused by a safety violation.

Each state views these claims slightly differently; yet in general, they are the same.

"Administrative functions of workers' compensation programs may be broadly divided into overseeing and/or disbursing funds and conducting hearings for the resolution of both disputed claims and insurance issues. Accordingly, bureaus, boards of agency appeals, or industrial commissions usually administer programs; and, in some instances, two state agencies handle functions separately: one agency to oversee and/or disburse, and the other to resolve conflicts.

"Notice of injury, death, occupational disease, or safety code violation usually must be given to either the employer or the state agency. In some states, notice is given separately from filing a claim; in others, filing a claim with a state agency or with an employer fills all the notice requirements of the statute.

"Evidence rules for administrative hearings are greatly modified from those applied in courts, and judicial review is often limited." [9]

Why Do We Need a Workers' Compensation System?

In essence, workers' compensation laws hold that industrial employers should assume costs of occupational disabilities without regard to any fault involved. Resulting economic losses are considered costs of production—chargeable, to the extent possible, as a price factor. The laws serve to relieve employers of liability from common-law suits involving negligence.

"Six basic objectives underlie workers' compensation laws:

1. Provide sure, prompt, and reasonable income and medical benefits to work-accident victims, or income benefits to their dependents regardless of fault;
2. Provide a single remedy and reduce court delays, costs, and workloads arising out of personal-injury litigation;
3. Relieve public and private charities of financial drains—incident to uncompensated industrial injuries;
4. Eliminate payment of fees to lawyers and witnesses as well as time-consuming trials and appeals;
5. Encourage maximum employer interest in safety and rehabilitation through an appropriate experience-rating mechanism; and
6. Promote frank study of causes of accidents (rather than concealment of fault)—reducing preventable accidents and human suffering." [10]

Have these laws done what they were expected to do and achieved results? The answer to this varies from state to state, and is dependent on many different things, including the person doing the evaluation.

In 1972, the National Commission on State Workers' Compensation Laws concluded that state laws were not doing what they were supposed to, and the Commission made 84 recommendations for the improvement of the system. They labeled 22.6 percent, or 19 of these, "essential." Notwithstanding the contrary assessment, the commission still felt that workers' compensation was a fundamentally sound system and of great value in our industrial economy.

The need for reform of state workers' compensation programs was also called for in January 1976, by the policy group of the Inter-Agency Workers' Compensation Task Force, with members from several U.S. government departments and agencies. In order to bring about a more effective management of workers' compensation, the Task Force suggested reform at the state level. They added that the federal government should monitor progress and provide technical assistance. The group also felt that the system of workers' compensation was on the brink of becoming more expensive, less effective, and less fair. After completing its mission, the Task Force was merged with the Division of State Workers' Compensation Standards in the Office of Workers' Compensation Programs, Department of Labor.

The Commission and the Task Force, by way of their statements showing the negative side of the situation, gave new incentive to the development and growth of workers' compensation laws.

While providing for injured workers is not argued, the problem has become the overwhelming costs involved in giving the benefits to the injured workers. In the last decade, workers' compensation costs have dramatically increased. The average cost of a claim is now around $34,000, which is nearly two times the amount of the average claim in 1980, according to the National Council on Compensation Insurance.

The real costs are not the only components that are rising in the system. The number of workdays lost has risen greatly, with the number of days lost per injury increasing. The basic breakdown of the costs is understandable and logical. Wage-loss and disability payments, added to the medical costs, are in part to blame. Medical benefits could represent nearly 40 percent of the total benefits paid.

Increases in medical costs are somewhat inherent in the system. Because medical bills are paid in full, workers do not have to share any of the burden of the costs; thus the worker sometimes will follow the doctor's recommendations without questioning whether he or she is receiving the best, most cost-effective treatment. This faith is fine as long as the doctor's treatment is right and fair.

Business people have to keep track of the claims, keeping in touch with both the doctors and the injured workers. The reason for this is that some doctors consider the workers' compensation system to be a cash cow. They look for any way possible to convert a treatable patient into a workers' compensation claim, thus allowing them to bill the system. This type of unscrupulous behavior causes everyone to lose, because this type of fraud simply and easily raises the costs all around.

While costs are rising, it needs to be clearly understood that the number of claims is also rising; it reached a ten-year peak in 1990. Litigation is another factor in the rise of claims and costs, which is ironic considering the system is supposed to be no-fault.

So now you have increasing costs, increasing numbers of injuries, and increasing types of injuries. In addition, not only are there dynamic injuries like sprains and strains, cuts, and falls, but there are also the newer types of injuries—those of a psychological or mental nature. Typically, these include stress or some form of abuse on the job. Also included in this list are any and all occupational diseases.

Here's an example of a physical injury and mental stress: a worker gets caught in a piece of machinery and loses a limb. This can and will cause any number of mental and psychological stresses that are work related. You also need to include here the other workers who happened to witness the limb being torn from the worker's body. This type of stress is very emotional and, of course, work related.

When cost factors of workers' compensation are looked at, two types of costs stand out. First, there are those that are direct, and second, those that are indirect.

The direct costs include medical benefits and compensation paid directly to the employee, and the costs of doing business, such as increases in workers' compensation insurance premiums.

"Indirect costs are the hidden costs of accidents that affect company earnings. Indirect costs include:

- Cost of time lost by the injured person;
- Cost of time lost by other employees who stopped to assist the injured worker or for other reasons;
- Cost of time lost by supervisors and managers assisting the injured worker, investigating the accident, arranging for another employee to cover the injured worker's responsibilities, and preparing accident reports or attending administrative hearings;
- Cost of damage to machinery, tools, or other materials;
- Cost of failure to produce or fill orders on time;
- Cost to employer under employee welfare benefit program;
- Cost to employer of light-duty work (where injured employee returns before full recovery and is paid even though he is not yet fully productive);
- Cost of overhead per injured employee (light, heat, air, etc.) which continues while the injured worker is nonproductive;
- Cost related to idle machinery; and
- Cost related to lowered morale following an accident." [11]

The following page contains a table of National Safety Council Estimates for the Cost of Worker Injuries and Illness in 1990 (see figure 1.5).

Any accident or work-related injury affects the company's bottom line. Direct and indirect costs are incurred. But what is the cost in terms of real dollars? The chart on page 15 (figure 1.6) shows the sales or services required to pay for a given accident based on the cost of the accident and the profit margin of the organization. For example, an accident that has a total cost of $50,000 will not be paid for until a company with a four-percent profit margin sells $1,250,000 worth of goods or services ($1,250,000 x .04 = $50,000). These dollar amounts present a convincing case that it is essential for companies to prevent and control workplace injuries.

Cost of Worker Injuries and Illness in 1990 (National Safety Council Estimates)

Direct (in billions)	
Wages Lost	$10.2
Insurance Administrative Costs	10.3
Medical Costs	8.7
	$29.2
Indirect (in billions)	
Money value of time lost by uninjured workers; cost of time to investigate accident; write report; etc.	$29.2
Damage to Tools, Equipment, and Products	5.4
	$34.6
Total	[$29.2 + $34.6] = $63.8
Cost Per Worker This figure indicates the value of goods or services each worker must produce to offset the cost of work injuries. It is not the average cost of a work injury.	$540
Cost Per Death	$730,000
Cost Per Disabling Injury	$23,000
Time Lost Because of Work Injuries (in days)	
Total Time Lost in 1990	75,000,000
Due to Accidents in 1990	35,000,000

Figure 1.5

Source: CCH, *Workers' Compensation Manual for Managers and Supervisors.* (Illinois: CCH Inc., 1992), 12.

SALES OR SERVICES REQUIRED TO PAY
FOR ACCIDENTS PROFIT MARGIN

ACCIDENT COSTS	1%	2%	3%	4%	5%
$ 1,000	$ 100,000	$ 50,000	$ 33,333	$ 25,000	$ 20,000
5,000	500,000	250,000	167,000	125,000	100,000
10,000	1,000,000	500,000	333,333	250,000	200,000
25,000	2,500,000	1,250,000	833,333	625,000	500,000
50,000	5,000,000	2,500,000	1,666,666	1,250,000	1,000,000
100,000	10,000,000	5,000,000	3,333,333	2,500,000	2,000,000
150,000	15,000,000	7,500,000	5,000,000	3,750,000	3,000,000
200,000	20,000,000	10,000,000	6,666,666	5,000,000	4,000,000
250,000	25,000,000	12,500,000	8,333,333	6,250,000	5,000,000
300,000	30,000.000	15,000,000	10,000,000	7,500,000	6,000,000
400,000	40,000,000	20,000,000	13,333,333	10,000,000	8,000,000
500,000	50,000,000	25,000,000	16,666,666	12,500,000	10,000,000

Figure 1.6
Source: CCH, *Workers' Compensation Manual for Managers and Supervisors*. (Illinois: CCH Inc., 1992), 13.

Reporting in the *Social Security Bulletin*, the U.S. Department of Health and Human Services estimated that employers spent over $53.1 billion to insure or self-insure their work-injury risks. This was $5.2 billion, or 10.8 percent, higher than the 1989 cost of workers' compensation. The prior year, the increase in cost was also 10.8 percent. The average cost per $100 of payroll was $2.36 in 1990, compared with $2.27 in 1989.

Medical costs totaled $15.2 billion in 1990. Compensation payments amounted to $23.1 billion, or about 60.3 percent, of all workers' compensation payments, which totaled $38.2 billion.

Occupational Safety and Health Administration (OSHA) Act of 1970

Until 1970, 90 million workers (now over 100 million) had no uniform and comprehensive provisions for their protection against workplace safety and health hazards. Each state had its own workers' compensation laws, yet the variances between states was great.

This act gave employees their first overall nationwide guidelines. When Congress was looking into this in 1970, they considered many different factors:

- Job-related deaths
- Nearly 2.5 million workers were disabled
- Ten times as many person-days were lost from job-related disabilities as from strikes
- Estimated new cases of occupational diseases totaled 300,000.

These factors put a great burden on the nation's businesses. The loss of production and wages, medical costs, and disability compensation were staggering the economy and business. Therefore, the Occupational Safety and Health Act of 1970 was passed by a bipartisan Congress "...to assure so far as possible every working man and woman in the Nation safe and healthful working conditions and to preserve our human resource."

Under the Act, the Occupational Safety and Health Administration was created within the Department of Labor to:

- Encourage employers and employees to reduce workplace hazards and to implement new safety and health programs or improve existing ones;
- Provide for research in occupational safety and health to develop innovative ways of dealing with occupational safety and health;
- Establish "separate but dependent responsibilities and rights" for employers and employees for the achievement of better safety and health conditions;
- Maintain a reporting and recordkeeping system to monitor job-related injuries and illnesses;
- Establish training programs to increase the number and competence of occupational safety and health personnel;
- Develop mandatory job safety and health standards and enforce them effectively; and
- Provide for the development, analysis, evaluation, and approval of state occupational safety and health programs.

While OSHA continually reviews and redefines specific standards and practices, its basic purposes remain constant. OSHA strives to implement its mandate fully and firmly with fairness to all concerned. In all of its procedures, from standards development through implementation and enforcement, OSHA guarantees employers and employees the right to be fully informed, to participate actively, and to appeal actions.

In general, coverage of the Act extends to all employers and their employees in the 50 states, the District of Columbia, Puerto Rico, and all other territories under federal government jurisdiction. Coverage is provided by the federal OSHA program or through OSHA-approved state programs.

The following are *not* covered under the Act:

- Self-employed persons;
- Farms at which only immediate members of the farm employer's family are employed; and
- Working conditions regulated by other federal agencies under other federal statutes.

What are the employers' responsibilities with regard to informing employees? The following is a list of documents that the employer needs to make accessible in the workplace.

- Job Safety and Health Protection workplace poster (OSHA 2203 or state equivalent) informing employees of their rights and responsibilities under the Act. Besides displaying the workplace poster, the employer must make available to employees, upon request, copies of the Act and copies of relevant OSHA rules and regulations. Any official edition of the poster is acceptable.
- Summaries of petitions for variances from standards or recordkeeping procedures.
- Copies of all OSHA citations for violations of standards. These must remain posted at or near the location of alleged violations for three days, or until the violations are corrected, whichever is longer.
- Log and Summary of Occupational Injuries and Illnesses (OSHA No. 200). The summary page of the log must be posted no later than February 1, and must remain in place until March 1.

With regard to the above, employees have the right to examine any records kept by their employers regarding exposure to hazardous materials, and the results of medical surveillance. In addition to access to records, employees also have the right to all information regarding any and all hazardous chemicals that may be in the workplace.

The OSHA Act has standards that fall into four major categories: general industry, maritime, construction, and agriculture. The *Federal Register* is one of the best sources of information on standards since all OSHA standards are published there when adopted, as are all amendments, corrections, insertions, or deletions. The *Federal Register* is available in many public libraries; annual subscriptions are available from the Superintendent of Documents, U.S. Government Printing Office, Washington, DC 20402.

When the workers' compensation system changed to a more no-fault system, it was done in order to keep down the effects of litigation. Workers who applied for benefits of workers' compensation only had to meet three different criteria:

1. There must be an injury or illness;
2. The injury or illness must have "arisen out of and in the course of employment"; and
3. There must be medical costs, rehabilitation costs, lost wages or disfigurement.

State workers' compensation benefits are supposed to replace two-thirds of the employee's lost income for a work-related injury or illness. Employers agreed to provide "no-fault" workers' compensation benefits as part of an historic compromise in which workers gave up their rights to sue employers. A worker, therefore, is not required to prove employer negligence, but is required to prove that an injury or illness was caused by the job.

Only a very small percentage—less than 10 percent—of those severely disabled from an occupational disease receive workers' compensation benefits. This lack of coverage is due, in part, to the difficulties involved in establishing the work-relatedness of the occupational illnesses because of (a) the length of time which has elapsed between the hazardous exposure and the onset of disability and/or death; (b) the multiple causes of diseases; and (c) associating specific firms.

Figure 1.7 on page 19 gives a very basic idea of how the reports get processed.

Summary

The history of workers' compensation has swung full arc, from total power in the hands of the employer/owner, to a system that offers some sense of economic protection to workers. Prior to the dawn of this century, a working man or woman had to accept whatever conditions, hours, and workplace dangers existed in his or her work. A worker absent from the job due to injury or workplace illness lost wages, and possibly the job itself. When the worker was injured (or worse), there was very little recourse through the courts to recoup lost wages or to pay medical bills. In 1908 in New York, only 12 out of 74 death cases resulted in compensation of $375-$3750 to the heirs, after the attorney was paid 25 percent; if the attorney took a fee of 33 1/3 percent, the families saw only $331.15-$3,311.50, less medical bills.

When the average American demanded better working conditions, a nationwide system evolved which offered income continuation, medical coverage, rehabilitation, and an opportunity to return to work when medically permitted. Though the nuances of the laws vary from state to state, the spirit of the laws—to protect the worker—is present. Parallel to the advent of workers' compensation, there is an increase in workplace safety: to keep workers' compensation premiums as low as possible, wise employers look for

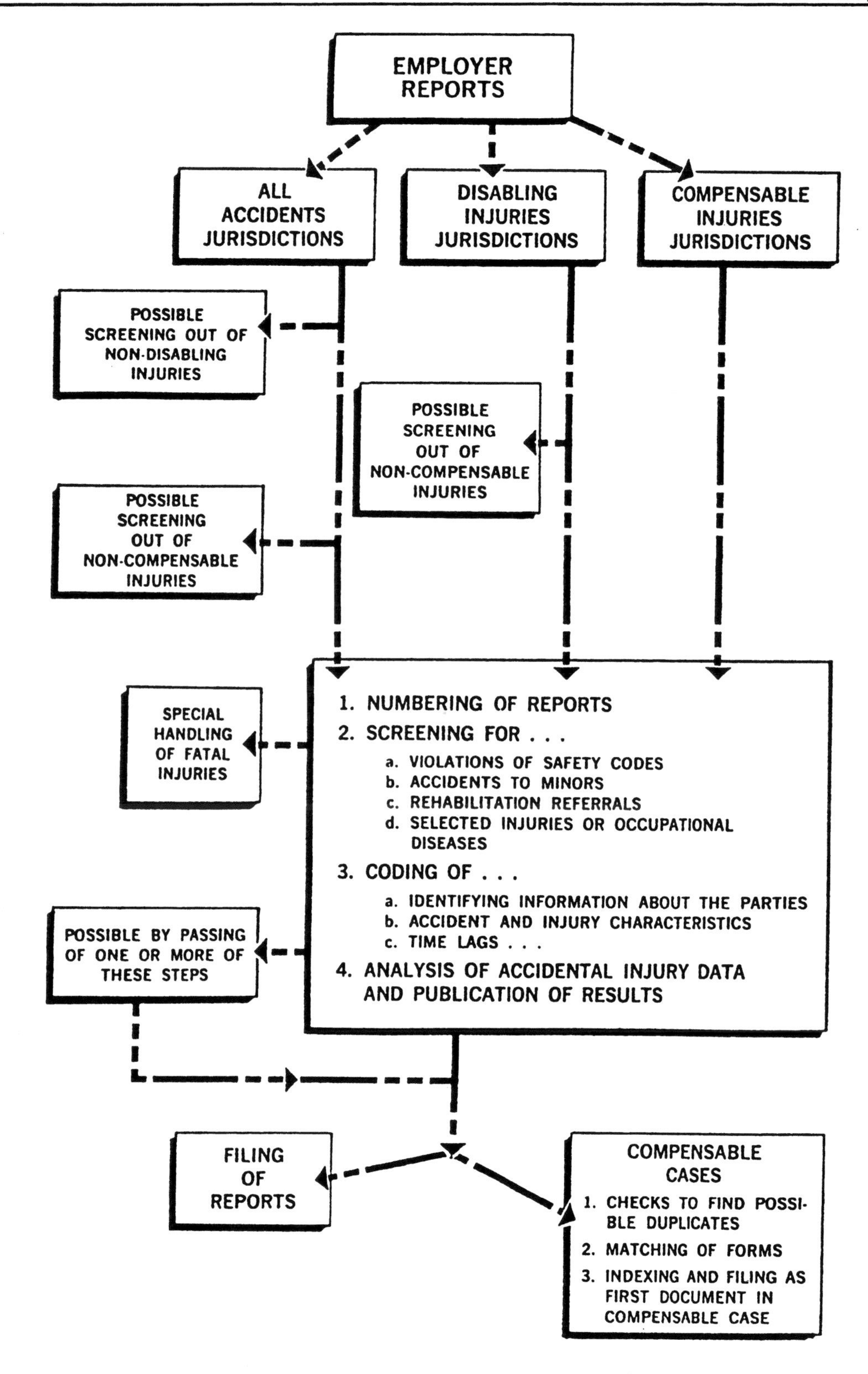

Figure 1.7: Processing First Reports of Work Injuries

Source: Jeffrey Nackley, *A Primer on Workers' Compensation.* (Washington, D.C.: Bureau of National Affairs, Inc., 1989).

every avenue to implement workplace safety measures. This is beneficial to both employers and employees.

If only it were a completely fair game. California, confronted by businesses fleeing to other states for radically lower workers' compensation premiums (due to runaway costs resulting from fraud, overbilling, and medically inappropriate treatments), overhauled its system. Previously, to prove a workplace stress claim, the worker had only to prove that 10 percent of the stress was work-induced. A bad hangnail might qualify! As of July 1993, the worker must now prove that at least 51 percent of the stress comes from the workplace. This one change has saved California's business community millions of dollars in nuisance claims. Other states are now examining the California example to see if it is applicable in their jurisdictions.

Pennsylvania presents its workers fewer hurdles to cross when proving a claim, while West Virginia offers the employers better protection. This has resulted in a two-tier system: workers try to press their cases in Pennsylvania while the employers, with post office box addresses over the border in West Virginia, claim they are West Virginia employers. Pennsylvania is revamping its workers' compensation system to adjust for this and other inequities.

While certain problems can be ironed out legislatively, others are solved on a personal basis. Open lines of communication between the employer, injured worker, and the treating doctors can only be maintained when the employer makes the effort to do so. By reassuring the injured worker that a job will be waiting for his return, that medical bills will be paid, and that compensation for lost wages will come regularly without the worker having to beg and plead, the employer will head off any looming clouds of suspicion in the employee's mind. Even though recovery is painful, the employee who knows that a roof will still be over his or her head and that food will still be on the table is much less likely to be persuaded to sue than the employee who feels injured, abandoned, and economically unprotected in a dangerous world. Employers who cut off payment of workers' medical bills or compensation pending claims handling greatly enhance the chances that they will be sued. Once lawyers are engaged on both sides, costs escalate.

Where does all the money from a settled workers' compensation claim go? The National Council on Compensation Insurance sites that an average claim is $34,000. Deduct medical benefits of 40 percent (or $13,600) and legal costs up to 33 1/3 percent (or up to $11,322), and our employee is left with $9078 to jingle in his or her pocket for pain and suffering (or more if the legal fees were less than average).

Chapter 2

Getting Started

Okay, it happened. Your employee comes to you to say that an injury took place. Now what?

This is when the entire workers' compensation process begins and where a large majority of the problems may begin.

Please understand that each state has its own idiosyncrasies with regard to what takes place after the injury has been reported. Since most of the forms used as examples in this book are from California, you will need to check with your state's workers' compensation board or insurance company. From either entity, you can get the exact forms needed. The basic process for filing the forms is similar nationwide.

Usually, there is a two-part form which both the employee and the employer complete. This form can be called "Employee's Claim Form" or something similar. The information requested on the top of this form is to be completed by the employee: the employee's name and address, area of body injured, when the injury happened, and a signature are required. (See figure 2.1 on the following page.)

The bottom of the form is to be completed by the employer. It includes the name of the employer, date injury was reported, date form was given to employee, name of insurance carrier, and signature of employer or his representative.

State of California
Department of Industrial Relations
DIVISION OF WORKERS' COMPENSATION

Estado de California
Departamento de Relaciones Industriales
DIVISION DE COMPENSACIÓN DEL TRABAJADOR

EMPLOYEE'S CLAIM FOR WORKERS' COMPENSATION BENEFITS

If you are injured or become ill because of your job, you are entitled to workers' compensation benefits.

Complete the **"Employee"** section and give the form to your employer. Keep the copy marked **"Employee's Temporary Receipt"** until you receive the dated copy from your employer. You may contact the State's Office of Benefit Assistance and Enforcement **at 1-800-736-7401** if you need help in filling out this form or in obtaining your benefits. An explanation of workers' compensation benefits is included on the reverse of this form.

You should also have received a pamphlet from your employer describing workers' compensation benefits and the procedures to obtain them.

RECLAMO DEL EMPLEADO PARA BENEFICIOS DE COMPENSACIÓN DEL TRABAJADOR

Si usted se ha leisonado o se ha enfermado en o a causa de su trabajo, Ud. tiene derecho de recibir beneficios de compensación del trabajador.

*Complete la sección **"Empleado"** y entregue el reclamo a su empleador. Quédese con la copia designada "Recibo Temporal del Empleado" hasta que Ud. reciba la copia fechada de su empleador. Si Ud. necesita ayuda para completar este reclamo o para obtener sus beneficios, póngase en contacto con la Oficina Estatal de Asistencia para Beneficios y Ejecución de la ley (Aplicación) llamando al **1-800-736-7401**. Al dorso de esta forma se encuentra una explicación de los beneficios de compensación del trabajador.*

Ud. también debería de haber recibido de parte de su empleador un folleto describiendo los beneficios de compensación del trabajador lesionado y el procedimiento para obtenelos.

Employee ***Empleado:***

1. Name *Nombre* ______ Today's Date *Fecha de hoy* ______
2. Home address *Dirección* ______
3. City *Ciudad* ______ State *Estado* ______ Zip *Código Postal* ______
4. Date of injury. *Fecha de la lesión (accidente).* ______ Time of injury *Hora en que ocurrió* ______ a.m. ______ p.m.
5. Address/place where injury happened. *Dirección/lugar dónde ocurrió el accidente* ______
6. Describe injury and part of body affected. *Describa la lesión y la parte del cuerpo afectada.* ______
7. Signature of employee. *Firma del empleado.* ______

Employer (complete this section and give the employee a copy immediately as a receipt):
Empleador: *(complete esta seccion y dele immediatamente una copia al empleado como recibo):*

8. Name of employer. *Nombre del empleador* ______

   Address *Dirección* ______
9. Date employer first knew of injury. *Fecha en que el empleador supo por primera vez de la lesión o accidente.* ______
10. Date claim form was provided to employee. *Fecha en que se le entregó al empleado el reclamo.* ______
11. Date employer received claim form. *Fecha en la que el empleado devolvió el reclamo completado al empleador.* ______
12. Name and address of insurance carrier or adjusting agency. *Nombre y dirección de la compañía de seguros o agencia administradora de seguros.* ______
13. Signature of employer representative. *Firma del Representante del Empleador.* ______
14. Title *Título* ______ 15. Telephone *Teléfono* ______

Employer: You are required to date this form and provide copies to your insurer and to the employee, dependent or representative who filed the claim within **one working day** of receipt of completed form from employee.

SIGNING THIS FORM IS NOT AN ADMISSION OF LIABILITY

Empleador: *Es requerido que Ud. feche este documento y que provea copias del mismo a su compañía de seguros y al empleado, representate o persona que dependa de él, que haya completado el reclamo, dentro de un día hábil después de haber recibido la solicitud completada de parte del empleado.*

EL FIRMAR ESTE DOCUMENTO NO SIGNIFICA ADMISION DE RESPONSABILIDAD

Original (Employer's Copy)
DWC Form 1 (REV. 7/90) 90 90691

ORIGINAL (Copia del Empleador
DWC Forma 1 (REV. 7/90

Figure 2.1: Sample of an Employee's Claim Form

State of California

EMPLOYER'S REPORT OF OCCUPATIONAL INJURY OR ILLNESS

Please complete in triplicate (type, if possible). Mail original and one copy to:

STATE COMPENSATION INSURANCE FUND

P.O. BOX 85488
SAN DIEGO, CA 92186-5488
Telephone: (619) 552-7100 Fax (619) 552-7110

OSHA Case No.

☐ Fatality

Any person who makes or causes to be made any knowingly false or fraudulent material statement or material representation for the purpose of obtaining or denying workers' compensation benefits or payments is guilty of a felony.

NOTICE: California law requires employers to report within five days of knowledge every occupational injury or illness which results in lost time beyond the date of the incident OR requires medical treatment beyond first aid. If an employee subsequently dies as a result of a previously reported injury or illness, the employer must file within five days of knowledge an amended report indicating death. In addition, every serious injury/illness, or death must be reported immediately by telephone or telegraph to the nearest office of the California Division of Occupational Safety and Health.

EMPLOYER

1. FIRM NAME — DIVISION — 1A. POLICY NUMBER

2. MAILING ADDRESS (Number and Street, City, ZIP) — 2A. PHONE NUMBER

3. LOCATION, IF DIFFERENT FROM MAILING ADDRESS (Number and Street, City, ZIP) — 3A. LOCATION CODE

4. NATURE OF BUSINESS, e.g., painting contractor, wholesale grocer, sawmill, hotel, etc. — 5. STATE UNEMPLOYMENT INSURANCE ACCT. NO.

6. TYPE OF EMPLOYER ☐ PRIVATE ☐ STATE ☐ CITY ☐ COUNTY ☐ SCHOOL DIST. ☐ OTHER GOVERNMENT - SPECIFY ____

EMPLOYEE

7. EMPLOYEE NAME — 8. SOCIAL SECURITY NUMBER — 9. DATE OF BIRTH (mm/dd/yy)

10. HOME ADDRESS (Number and Street, City, ZIP) — 10A. PHONE NUMBER

11. SEX ☐ MALE ☐ FEMALE — 12. OCCUPATION (Regular job title—No initials, abbreviations or numbers) — 13. DATE OF HIRE (mm/dd/yy)

14. EMPLOYEE USUALLY WORKS ____ hours per day ____ days per week ____ total weekly hours — 14A. EMPLOYMENT STATUS (See instructions in 14A continued below.) ____ regular full-time ____ part-time ____ temporary ____ seasonal — 14B. Under what class code of your policy were wages assigned?

15. GROSS WAGES/SALARY $____ per ____ — 16. OTHER PAYMENTS NOT REPORTED AS WAGES/SALARY (e.g., tips, meals, lodging, overtime, bonuses, etc.)? ☐ YES, $____ per ____ ☐ NO

INJURY OR ILLNESS

17. DATE OF INJURY OR ONSET OF ILLNESS (mm/dd/yy) — 18. TIME INJURY/ILLNESS OCCURRED ____ A.M. ____ P.M. — 19. TIME EMPLOYEE BEGAN WORK ____ A.M. ____ P.M. — 20. IF EMPLOYEE DIED, DATE OF DEATH (mm/dd/yy)

21. UNABLE TO WORK FOR AT LEAST ONE FULL DAY AFTER DATE OF INJURY? ☐ YES ☐ NO — 22. DATE LAST WORKED (mm/dd/yy) — 23. DATE RETURNED TO WORK (mm/dd/yy) — 24. IF STILL OFF WORK, CHECK THIS BOX ☐

25. PAID FULL WAGES FOR DAY OF INJURY OR LAST DAY WORKED? ☐ YES ☐ NO — 26. SALARY BEING CONTINUED? ☐ YES ☐ NO — 27. DATE OF EMPLOYER'S KNOWLEDGE/NOTICE OF INJURY/ILLNESS (mm/dd/yy) — 28. DATE EMPLOYEE WAS PROVIDED EMPLOYEE CLAIM FORM (mm/dd/yy)

29. SPECIFIC INJURY/ILLNESS AND PART OF BODY AFFECTED, MEDICAL DIAGNOSIS, if available, e.g., second degree burns on right arm, tendonitis of left elbow, lead poisoning.

30. LOCATION WHERE EVENT OR EXPOSURE OCCURRED (Number, Street, City) — 30A. COUNTY — 30B. ON EMPLOYER'S PREMISES? ☐ YES ☐ NO

31. DEPARTMENT WHERE EVENT OR EXPOSURE OCCURRED, e.g., shipping department, machine shop. — 32. OTHER WORKERS INJURED/ILL IN THIS EVENT? ☐ YES ☐ NO

33. EQUIPMENT, MATERIALS AND CHEMICALS THE EMPLOYEE WAS USING WHEN EVENT OR EXPOSURE OCCURRED, e.g., acetylene, welding torch, farm tractor, scaffold.

34. SPECIFIC ACTIVITY THE EMPLOYEE WAS PERFORMING WHEN EVENT OR EXPOSURE OCCURRED, e.g., welding seams of metal forms, loading boxes onto truck.

35. HOW INJURY/ILLNESS OCCURRED. DESCRIBE SEQUENCE OF EVENTS. SPECIFY OBJECT OR EXPOSURE WHICH DIRECTLY PRODUCED THE INJURY/ILLNESS, e.g., worker stepped back to inspect work and slipped on scrap material. As he fell, he brushed against fresh weld, and burned right hand. USE SEPARATE SHEET IF NECESSARY.

36. NAME AND ADDRESS OF PHYSICIAN (Number and Street, City, ZIP) — 36A. PHONE NUMBER

37. IF HOSPITALIZED AS AN INPATIENT, NAME AND ADDRESS OF HOSPITAL (Number and Street, City, ZIP) — 37A. PHONE NUMBER

38. WAS ANOTHER PERSON RESPONSIBLE? ☐ YES ☐ NO — 39. WAS INJURED AN EXECUTIVE OFFICER OR A PARTNER? ☐ YES ☐ NO

14A. EMPLOYMENT STATUS CONT. (Check current status of employment, not status at time of injury.)
____ UNEMPLOYED ____ ON STRIKE ____ DISABLED ____ RETIRED ____ LAID OFF ____ OTHER

Completed by (type or print) — Signature — Title — Date

DO NOT USE THIS COLUMN: Case No. / Ownership / Industry / Occupation / Sex / Age / Daily hours / Days per week / Weekly hours / Weekly wage / County / Nature of injury / Part of body / Source / Event / Sec. Source / Extent of Injury

SCIF 3067 (REV. 2-93) FILING OF THIS REPORT IS NOT AN ADMISSION OF LIABILITY. A CLAIM FORM MUST BE GIVEN TO THE INJURED WORKER WITHIN ONE WORKING DAY OF YOUR KNOWLEDGE OF OCCUPATIONAL INJURY OR ILLNESS WHICH RESULTS IN LOST TIME OR MEDICAL TREATMENT.

Figure 2.2: Sample of Employer's Report of Occupational Injury or Illness

Once completed, this form is sent to the insurance company so that it can begin to process the claim and set up a file. Copies of this form should be kept on file because this form is usually needed by doctors, lawyers, and even the Workers' Compensation Appeals Board.

This form generally needs to be completed within 1-3 days; otherwise, some states can fine the employer for not carrying out his/her obligations.

Now comes the next form, typically called "Employer's Report of Occupational Injury or Illness" (figure 2.2 on previous page). This form is filled out exclusively by the employer. It gives the basics of what happened, who was injured, and the employer's information. This form goes to the insurance carrier and employer only.

After this has been completed, the employer has the option to (1) send the employee to the doctor of the employer's choice; or (2) let the employee go to the doctor of his/her choice.

In California, the employee has the right to choose a predesignated physician, and the employer is obligated to agree. The only restriction to this is that the predesignation must be done in writing prior to the injury.

This step, which seems relatively simple, can start a whole negative process in the employee's mind. If the employee is forced to go to the "company doctor" and receives poor treatment, the employee may perceive that the doctor is siding with management. If the doctor returns the employee to work while the symptoms or the pain still persists, the employee may really get mad. This episode could generate real problems down the road. (If the doctor's interpersonal skills are poor, you may wish to consider a new health care provider.)

At the point when the employee has notified management, has consulted a treating doctor, and has filled out appropriate paperwork, most employers relax. They figure that all is handled and they can go back to doing their regular routine, forgetting about this entire incident.

This is a major mistake for an employer or manager to make. The employer still needs to be aware of what is going on. Remember, it is your employee, your company—you'd better know what is going on.

What if this employee decides that returning to work is impossible because of continued back pain? You may have to replace a good, competent employee.

Employees may also decide to go the other way and milk the system, figuring they are getting disability checks for approximately 75 percent of their gross salary every week while not working. Why should they go back to work? This way they can get a short or long paid vacation.

Now you are on the hook. You have a business to run, work to be done, and you may have to run your company an employee short for a while. Does this become a burden? You bet.

One thing that is important to remember here is that the larger the company, the bigger the workers' compensation insurance bills and the greater the overall costs. Granted, a small business might severely feel the effects of having an injured employee out for two weeks but, more importantly, it has just as much to gain by controlling costs/job time lost as the megacorporation.

Can management do something about it? Absolutely. You need to keep track of what is going on, contact the employee on a regular basis, contact the doctor on a regular basis, and even speak to the insurance company to find out what is going on.

As the employer, your responsibility really starts when you begin your business. It is up to you to find a reliable, cost-effective medical or chiropractic office to work with.

Again, this seems like a very simple decision. In reality, it may not be. Classically, most businesses simply locate the nearest Urgent Care facility and funnel all their injured employees through its front doors. This is all well and good when the injury requires drugs, surgery, or stitches. But if the injury is a sprain or strain, the Urgent Care route may be neither the most cost-effective, nor the quickest healing treatment.

It has been found that if companies improve relations with their employees, they are able to reduce their workers' compensation costs.

As the employer, it is sometimes difficult to put yourself in the shoes of your injured employee. Many different thoughts, be they confused, scared, or complicated, go through the employee's mind within the first few days following an injury.

Some of those questions and thoughts may include the following:

- Where do I go for medical care?
- Will I still get paid if off work?
- Will my job be there when I get back to work?
- What if I can't do the same job after recovering from my injury? Will my employer have some other job for me?
- Who should I talk to about this? Where is the best place to get information regarding my injury?
- I keep getting calls from a *claims adjuster*. Who is this person?
- Does anybody really care about me? Is anybody on my side?
- Do I need an attorney?

Employers need to address these issues. Not only does this relieve the employee's fears, but it can also help reduce costs. If these concerns are not addressed, it can set up a very negative cycle between management and employees.

Remember, the employee is injured, scared, and in pain. He or she probably has no idea what is going on and just wants to be treated fairly and with respect.

You, on the other hand, are not in pain and have no injury, yet you have a business to run, so worrying about a hurt employee may not be on the top of your list of priorities. It doesn't have to ruin your whole day, destroy your business, or make you lose sleep at night. What it does mean is that you have a responsibility to that injured worker to handle his needs and answer his concerns.

This does two things. First, it eases the workers' fears and concerns regarding pay, job status, and security. Secondly, by showing a concern for the worker, it keeps morale up, makes the employee feel that he or she is valued, and helps to eliminate the adversarial effects that can come up in a workers' compensation case.

If you, as the employer, along with managers and supervisors, help the employee through the maze of workers' compensation, then the possibility of the injury turning into a nightmare is mitigated.

Steps to Take After an Injury

Under normal circumstances, the injured worker is the first person to know about the injury, with a supervisor or manager alerted soon after that. It is vitally important to be supportive to the employee, even if you might think the injury is suspect.

Look at all injuries as being legitimate. The simplicity of this statement is apparent, yet the ability to do it is definitely not easy. However, if you and your managers treat the injured worker well and with respect, the chances of messy litigation are much lower.

1. A manager or supervisor must respond to the worker. Get all the facts regarding the injury and let the employee know that there are systems in place to handle injuries within the workplace. As the manager, it is necessary for you to stop what you are doing and help the worker. Make the worker feel comfortable and secure and reduce his or her fears as much as possible.

2. If first aid is needed and can be accomplished on the job site, do it. If the injury is beyond first aid, get the worker to medical care immediately. This is the point where injury management first takes place. No matter how minor the injury might be, handle it as though it was a critical emergency. Have someone, including yourself, ready to

transport the employee if necessary. These kinds of arrangements should be handled in advance, so that when an injury happens, chaos does not break out.

3. After the preliminary medical care/first aid, the next step is to begin to document the details of the injury. Most states have accident forms that need to be filled out by both the employee and the employer. You might want to have your company make up a form of its own, something simple and easy, that can provide the necessary details, and can again show the employees that their company cares about them.

4. It is at this time that a more detailed explanation of the company's workers' compensation procedures needs to be explained, either to the employee or to the family of the injured worker. This explanation needs to include when payments will be made, how additional medical care is handled, basics on disability leave, rehabilitation, etc.

Again, you need to present this information about compensation and job security clearly and calmly to the employee and the family. The worker's guard is up, fears are at their peak, and he or she has no idea how the system will work. Reassure the worker and the spouse that they will be taken care of and that the worker is not merely a number at the company. The importance of this cannot be stressed enough.

At this time, it is also important for the supervisor or manager to serve as the intermediary between the company and the injured worker.

- Show workers and their families that you care;
- Do everything you can to allay the employee's fears regarding lost wages, job security, seniority, and other issues;
- Try to let the employee know that everyone benefits from his return to work.

There is an important distinction between (1) pushing a not-yet-recovered employee back to work prematurely and (2) making it clear that you genuinely want a fully healed employee back at the right time. Let the injured worker know the welcome mat is out for him.

If your company drops the ball at this point by stalling payments to the employee, or not giving the employee the time off to heal, or fails to reassure the employee that all will be well, you can expect the employee will seek solace in the office of an attorney. This opens up a very large can of worms.

5. The supervisor or manager needs to *stay involved.* Since medical care is where a large percentage (up to 40 percent) of workers' compensation dollars go, keep a good eye on what is happening. Follow up with the health care provider as to the status of the

worker. If you cannot directly communicate with the doctor, talk to the worker involved. See if the worker is getting better, what tests may be planned for, and generally how he or she feels about the health care being received. If the employee indicates that he or she is dissatisfied with the care being received, ask where the employee might want to seek care. Volunteer to find referrals for the employee. (Good referrals are your Rolodex, fellow employees, and referrals from other doctors.) Providing the employee with the names and phone numbers of three potential doctors drives home the message that you really care if the person gets better.

Positive communication with the injured worker is strongly stressed. It is of utmost importance to keep a lid on costs, both direct and indirect.

6. It is important to know who your medical provider is. To look for a doctor after the injury has happened is too late. It needs to be done as soon as you go into business. You need the names of health care providers with whom you have a good rapport—regardless of whether the doctors are medical, osteopath, chiropractic or something else. If you have not established a relationship with a doctor, chiropractor, osteopath, etc., do it today. After an employee has been injured is not the time to find the appropriate kind of doctor for the injured worker. You need to do this before any and all injuries happen.

Don't just go to the closest doctor's office. Talk to a number of different types of doctors to find out who they are, what their basic procedures are in a workers' compensation case, and whether they can really accommodate your needs. It is also wise to have more than one facility for use at any given time. Have a medical facility for those injuries that require medical care and a chiropractic office to handle sprains and strains (the number one type of work injury).

Different states have statutes to determine who has the right to tell the employee which medical facility to go to. Here again is a place where communication between the company and the employee can be critical. As the employer, you can always suggest where the employee should be seen for care, but if the employee requests a particular doctor, give him the option of seeing his regular doctor, instead of some unknown, company doctor.

In all cases of this nature, try and put yourself in the employees' shoes. As the manager or supervisor, depending on the size of your company, run mock injuries with the other management people so they can see what it is like to be an injured worker. Treat the manager just like you would treat a worker and see what happens.

7. One thing that experience and research has shown is that some employers try and get around filing an injury claim. On a very minor injury, it really isn't worth the time to fill out all those forms—or so you may think.

If you talk to others in management, you will hear many horror stories about the minor injury they did not report—and how it backfired on them. The minor problem really turned out to be a major incident. The employee now has an attorney, your company is getting fined for not filing a report, your workers' compensation insurance premiums just went up, etc. Just file the claim no matter how minor it seems, no matter how small or incidental it may be.

8. After you have filled out the accident reports, gotten care for the employee, and resumed your usual job duties, now is the time to begin a basic investigation of the accident.

Your purpose here is not to undermine the injury. It is threefold: (1) determine how the injury happened so that you are able to remove those causes; (2) after finding the cause, determine how to prevent it from happening again; and (3) refocus your management team's attention to safety.

Remember, a large percentage of injuries can be prevented. If during the usual course of the day, employees and management observe unsafe conditions, such as incorrect lifting methods, now is the time to correct it. Work in a manner most conducive to NOT getting hurt.

9. Reassure the employee that it is important for him to return to work as soon as possible. Let him know that there will always be a job for him upon his return, whether it is regular, modified, or restricted duty, as decreed by the treating doctor. As soon as permissible, the employee will resume the pre-injury job with its rights and benefits.

Communicate clearly and frequently with the medical provider. Early in the regimen of care, give the doctor a basic job description. If the doctor understands what the employee/patient needs to be capable of doing, therapy can be focused, and the patient can be returned to modified or regular duty earlier. The triad of management, treating doctor, and injured employee can work together to restore a renewed sense of health for that employee.

10. Maintain open lines of communication with the injured worker—something you should have been doing all along. A marriage fails when partners don't communicate—and ends up in the hands of an attorney. A workers' compensation case in which management doesn't communicate positively and effectively with the injured worker can also end up in the hands of an attorney. Let the employee know that the company cares about him or her and really wants the worker back.

When an employee or injured worker gets in the mindset of being a "worker on disability," it can create an attitude which argues, "why should I bother going back to work? I am being paid 75 percent of my gross salary tax-free, not working, and getting free medical care. What could be better?" This obviously is a very dangerous viewpoint for a worker to take. Once workers have no desire to return to work, they usually don't. Huge medical and rehabilitation bills are run up, compounded by disability payments and other costs—all of which show up as higher workers' compensation premiums for the company.

The employee has responsibilities here also. He or she cannot just get injured and decide to tell you about it three or four weeks after it happened. In most states, the injured employee is required to give the employer notice of the injury. This requirement is done as much for the employee's benefit as it is for the employer's.

Some states have certain time limit requirements for filing a claim. These time limits have caused problems in some states because certain types of work-related illnesses do not show up for many years after exposure, such as radiation poisoning from working with atomic energy.

Most states have a 30-day statute for filing a claim. This is 30 days from the time the employee first becomes aware of an injury. When the nature of the injury or disease or its relationship to the employment is not known to the employee, then the time for filing the claim is not to begin until (1) the employee knows, or with the exercise of reasonable diligence should know, the existence of the injury and its possible relationship to his employment; and (2) he sustains disability, or incurs a scheduled loss.

All jurisdictions require employers to make some report of injuries. Although there are variations, in general workmens' compensation jurisdictions can be classified into three categories.

The first category includes jurisdictions in which employers are required to report "all injuries." This may not be further defined and may be variously interpreted as either excluding, or not excluding, first aid cases. Technically, this is the broadest of all classifications, except that its lack of definition may mean that it is widely disregarded in practice.

The second category is when the employer is required to report work injuries which result in (1) time loss beyond the shift or the working day in which the accident or injury occurred; or (2) a situation where permanent disability occurs.

The third category includes only those accidental injuries that result in permanent disability, or loss of time beyond the waiting period specified in that state. These are the potentially reportable cases.

Summary

This area in the workers' compensation system is where it all begins. Getting things properly handled at the beginning helps to make the process simpler, less stressful, and much more pleasant for the employee and your business. Report injuries, do the proper paperwork, and keep involved with the employee to keep the process moving along. Cost containment has become the "buzzword" of the 90s and in the workplace it might be more critical than the home. By following some simple steps in the beginning, before the injury, you will be able to cut costs in the realm of workers' compensation.

Chapter 3

Injuries and the Injured Worker

What constitutes an injured worker?

This question is both very simple and very difficult to answer. On the one hand, an injured worker is someone who gets hurt at work. Sounds simple.

Well, is an injured worker someone who only gets injured on the job site? Is an injured worker someone who gets injured while driving his/her own car on company time? How about the person bending over in the parking lot after work who hurts his back? What about someone who gets harassed at work? How about the employee joking around with his fellow employees who gets his hand stuck in a piece of machinery? How about the executive who has a heart attack in the workplace?

What about those diseases that we hear about: black lung, asbestosis, carpal tunnel, tendonitis?

These are just a few of the areas that affect every one of us, as future employees and employers. For your information, all of the above could be legitimate work-related injuries. While "occupational disease" and "work injury" are considered to be synonymous terms, in reality, they are not.

The definition of "occupational disease" varies a great deal between different states and different jurisdictions. Commonly, the term means a prolonged exposure to the specific work environment or work type, and thus becomes somewhat undifferentiable from an injury. In other jurisdictions, the term is limited to certain specific listed injuries, which are considered to be caused by specific jobs or job descriptions. The distinctions and definitions are also different from accidents or injuries because they cannot be said to be unexpected, as accidents are.

It is best for most purposes, however, to apply a common sense definition to the term; that is, to think of occupational diseases as those ailments that are associated with particular industrial trades or processes. Obvious examples include coal miners' pneumoconiosis or "black lung" disease, radiation illness, silicosis, and the asbestos-related diseases notoriously associated with the insulation industry. Not-so-obvious examples would include arthritis caused by repetitive trauma, vibrations associated with some physical labor, mental-stress ailments associated with high-pressure positions, and perhaps, certain cardiovascular problems sometimes said to be connected with police and firefighting positions.

One way to arrive at a very good description of the distinction between injury and occupational disease is to review literature on the subject. Arthur Larson, an authority in workers' compensation, gives the following description: "The important boundary becomes now, not that [boundary] separating occupational disease from accident, since compensability lies on both sides of that boundary, but the boundary separating occupational disease from diseases that are neither accidental nor occupational, but common to mankind and not distinctively associated with employment."

It needs to be stated here that virtually any, and all, diseases originating in the workplace can and usually are recognized as occupational diseases.

An injury does not have to be the obvious thing we all think of, such as hurting the back by lifting a large, very heavy crate. The injury can be very subtle or very obvious. So this simple little question, "What is an injured worker?", takes on many twists and turns, depending on the circumstances.

Injuries happen every day in the work environment. Some of them are blatant, while the largest growing portion of injuries tends to be more of a cumulative trauma nature.

The blatant injury is pretty obvious: lifting something too heavy or in a manner that is not proper to the body, falling from a roof or scaffolding, cutting a part of the body, etc.

Cumulative trauma is a very different animal. It happens over a long period of time, generally from a highly repetitive movement. No signs or symptoms may show up for years after the injury begins to happen. It is quite common for someone to leave a labor-intensive, highly repetitive type of job and then months later have symptoms show up for no apparent reason. Patients with carpal tunnel, which occurs in the wrist, may show no symptoms while they are engaged in the work which is actually causing the injury. Lo and behold, several months after the employee leaves the job for something totally unrelated, the typical symptom pattern appears, seemingly out of nowhere.

An injured worker is defined by California and most states as someone who has experienced any injury or disease resulting from employment, injuries to artificial members, dentures, hearing aids, eyeglasses, and medical braces of all types.

A definition of specific or cumulative injuries is as follows: an injury which may be either (a) "specific," one that occurs by one incident or exposure which causes disability or need for medical treatment; or (b) "cumulative," one that occurs as a repetitive mental or physical trauma extending over a period of time, where the combination of effects causes a disability or need for medical treatment.

Stress-related injuries are a highly contested, highly controversial area of workers' compensation. Each state defines this differently. Until July 1993, California defined a stress-related work injury as one that had a minimum of 10 percent of the stress caused by the workplace. The percentages were just changed to 51 percent of the stress taking place in the workplace.

Montana has an interesting definition for accident or injury. It is "a tangible happening of a traumatic nature from an unexpected cause or unusual strain resulting in either external or internal physical harm, and such physical condition, as a result therefrom and excluding, disease not traceable to injury." This is a very complex definition for the more standard idea of what a covered accident claim is.

At this point, it is important to look at the different types of injuries. The purpose of the following information is to go into more detail, without sounding like a lawyer or a doctor.

"In the Course of" and "Arising Out of" Employment

When reviewing information on workers' compensation, there are two terms that come up over and over again: "in the course of" and "arising out of" employment. These terms are used by most states to determine work-relatedness of an injury or illness.

"Some states, like Utah, changed it to 'arising out of *or* in the course of employment.' The interpretation of this term or terms is very much in debate. In a great many cases you will hear these terms split in two, with 'arising out of the employment' [referring to] casual origin, and 'course of employment' [referring] to the time, place and circumstances of the accident in relation to the employment.

"When an injury is found to be 'in the course of and arising out of employment,' it [is] then said, by definition, to be work-related. Generally, it means that the terms indicate that the injury was caused by the work. This determines and defines that the injury has compensation factors.

"The quite logical suggestion has therefore been made that use of the terms 'in the course of and arising out of employment' only add confusion to the law, and that the question should be, and usually has been, whether or not an injury, disease, or bodily condition was work-related, i.e. whether it was caused by the employment." [12]

So what does all this mean? Generally, the terms "in the course of" and "arising out of" employment mean that the injury was caused by work; these terms also determine and define when the injury has compensation factors. Legally, most courts have interpreted this to mean that the injury was caused by an increased risk to the worker by way of his employment.

The terms "in the course of and arising out of employment" are often treated as though they were one relatively simple concept. The problem occurs when situations arise that need to be reviewed by courts or administrative judges to determine the separateness of these two statements.

In the Course of Employment

An injury is said to arise "in the course of employment" when it takes place during the time of employment, at a place where the employee may be, and while the employee is performing his or her usual and customary duties or is engaged in doing something incidental thereto.

Most people misconstrue the meaning of the statement "in the course of employment" to be that the employee had to be in the course of his or her employment; what it really means is that the *injury* has to come about in the course of employment. The following are categories of claims where injuries occur "in the course of employment":

- Employees who are engaged in some errand or activity not directly connected with work but that gives some subsidiary benefit to employment—for example, while taking a class or while participating in sporting events—for the purpose of increasing their value as employees. It is difficult to fashion a general rule in these cases. Differing results have been reached in similar circumstances. The New Jersey Supreme Court in *Strezelecki v. Johns-Manville Products Corp.*, for example, allowed a claim for a death that occurred when an employee was involved in an automobile accident on his way to study for university classes that his employer paid for and that were in furtherance of his career as a financial analyst in his employment. On the other hand, the Florida Supreme Court in *Mathias v. South Daytona* disallowed a claim for an injury sustained by a police officer during a softball game in which he was strongly encouraged to participate. There is no good rule of thumb for making sense out of the varying results in this category of injuries.

- Employees at an employment-sponsored function away from the regular location of work—at a picnic, dinner, party, etc. These injuries are generally recognized as being in the course of employment and compensable, particularly if the employee was compelled, either directly or indirectly, to attend. The dispositive question in such cases of whether the employer derived some benefit from the affair is usually answered in the affirmative; the Rhode Island Supreme Court's decision of *Beauchesne v. David London & Co.*, allowing a claim of an employee who sustained a severe injury when he became drunk at a company Christmas party, cannot really be considered aberrant.

  On those occasions when a claim in this category is disallowed, the reason usually given is that it was not clear whether the employer actually sponsored the event, as in *Chilton v. Gray School of Medicine* (N.C.); or that the employer's sponsorship was so minimal that it received no substantial benefit from the function, as in *Pasko v. Beechef Co.* (Minn.).

  Nevada's workers' compensation law specifically excludes injuries sustained at a social gathering unless the workers are paid, while the California code states that voluntary participation in off-duty sports events is not covered.

- Employees such as traveling salesmen, who have no fixed location of employment, or workers who operate from their homes. The general rule in such cases is easy: injuries are compensable whenever the individual is performing work for the employer. It is more difficult to determine when the work begins and when it ends, in particular situations.

- Employees whose work requires some entertainment of customers or clients, so that the distinction between location of work and site of personal activities is unclear. Again, the general rule is deceptively simple: employees engaged in work for the employer are "in the course of" their employment. Courts generally allow compensation if there is a sufficiently close connection between the entertaining and the employee's work (*Shunk v. Gulf American Land Corp.* (Fla.), *Green v. Heard Motor Co.* (La.), *Harrison v. Stratton* (N.J.)). But the factual inquiries into this question are often extremely difficult, and courts often resort to the liberal construction of workers' compensation law as an aid in their resolution.

- Employees engaged in some personal activity while on the employer's premises during work hours. The central question in such cases is usually phrased: was the act purely in the employee's interest or was it incidental to employment? But divergent results have been reached on the same facts. Some indication of the varying views possible in these cases was expressed in *Leckie v. H.D. Foote Lumber Co.* by a Louisiana appeals court that, after reconsideration, reversed itself and denied compensation to a lumber-mill employee who was injured while sawing a piece of scrap lumber for his personal use.

  Relevant factors in resolving this issue include whether the injury took place during regular work hours, whether stand-by or idle time is part of the nature of the job, and whether the employee's engaging in such personal activity was known by the employer and whether it was permitted or prohibited.

  Factual variations are so wide in this category that only a few examples will be given. The North Carolina Supreme Court denied compensation to a watchman for an injury he sustained while washing his car *(Bell v. Dewind Bros., Inc.)*; a department-store employee was held by the Ohio Supreme Court not to be in the course of employment while making a purchase of articles for himself "on company time" *(Industrial Commission of Ohio v. Ahern)*; a service-station operator who was injured while repairing his own car at work was granted compensation by the Nebraska Supreme Court *(Chrisman v. Farmers Cooperative Association)*; a laundry employee's injuries sustained while pressing a skirt for a coworker after working hours were held to be not compensable by the Tennessee Supreme Court *(Hinton Laundry Co. v. De Lozier)*.

At this point, it should become clearer that asking whether an injury was sustained "in the course of employment" is simply one aspect of the question of whether an injury was caused by employment. In those categories of factual situations discussed above, it is easier to approach that question by asking whether the injury was sustained "in the course of employment."

Arising Out of Employment

It is a relatively arbitrary decision to analyze one fact category—for example, injuries sustained while an employee is engaged in personal activity during business hours under the "in the course of employment" approach rather than the "arising out of employment" language. Under the former, one would ask whether the personal activity removed the worker from "the course of employment." Under the latter, one would ask whether the injury arose out of the individual's work rather than out of his personal activity. For the basic question, "Was the injury caused by the work?", the answer should be the same in both scenarios.

"The following categories of claims illustrate this point. The usefulness—let alone the analytical validity—of distinguishing 'course of employment' from 'arising out of employment' more or less disintegrates, and it is best in these 'arising out of employment' cases to keep only the ultimate question in mind: Did the employment cause the injury?

- "Employees who are away from the site of employment but are on standby or on call for work. The general rule is that if an individual is injured performing a duty of work or is injured because of a risk incurred because of the job, the injury is compensable. An on-call management employee who was ordered by his employer to keep the plant in operation during a labor strike and who was beaten and called a 'dirty scab' when he stepped out of his car at his residence was held to have suffered a compensable injury in the New Jersey case of *Meo v. Commercial Can Corp*. Some cases, like *Haugen v. State Accident Insurance Fund* (Or.) and *Kelly v. County of Nassau* (N.Y.), put the rule conversely: Injuries sustained while performing work voluntarily or for the benefit of the employee, rather than of the employer, are not compensable.

- "Individuals who are required to live on the employer's premises and [are] injured while performing some personal activity. The general rule is that injuries are compensable if incurred in the performance of work duties or [are] reasonably attributable or incidental to the conditions under which such an employee lives. Exceptions to this general rule of compensability are usually explained by an employee's not being required to live on employer's premises but living there voluntarily; as in, for example, *Guiliano v. Daniel O'Connell's Sons* (Conn.), *Kraft v. West Hotel Co.* (Iowa), and *Guastelo v. Michigan C.R. Co.* (Mich.).

- "Workers who are on break on or off the employer's premises during the normal workday. Injuries occurring during rest or eating periods at an authorized place provided by the employer are compensable. Most courts have held that if the meal is taken off premises, resulting injuries are not compensable. But a minority have decided to the contrary; as for example, the courts in *Littlefield v. Pillsbury Co.* (Ohio) (employee returning from break [was] injured in a motor vehicle accident just outside the entrance to employer's premises) and [in] *Hornyak v. Great Atlantic & Pacific Tea Co.* (N.J.). Jurisdictions are about evenly split on the compensability of injuries sustained during breaks on the employer's premises but away from authorized lunch or recreation areas. In particular fact patterns, findings of compensability have been premised on determination that the employee was exposed to a special hazard by the nature of employment (e.g., a taxi driver forced to eat in cafeterias, *Relkin v. National Transp. Co.* (N.Y.)).

- "Employees who are injured during work as a result of horseplay or quarrels with co-workers. These employees are covered by workers' compensation if they did not initiate the activity but were only the innocent victims. There is a trend toward liberalization of the general rule if the horseplay is an expected part of the employment (e.g., messenger boys shooting paper clips from rubber bands, *Johnson v. Loew's* (N.Y.)), but a small number of decisions do not appear to recognize claims for *any* injuries resulting from horseplay (e.g., the Georgia case of *Kight v. Liberty Mutual Ins. Co.).*

- "Employees who violate work rules and are injured. Misconduct of this sort does not render injuries noncompensable unless the rule that is violated is so basic to the job that the worker is removed from the sphere of employment, as in the case of South Carolina police chief who, in violation of work rules, rode in a fire truck and was thereby injured *(Black v. Town of Springfield).*" [13]

Going and Coming

Another area that is necessary to cover here is that of the rule known as the "going to and coming from" rule. For employees who have fixed hours and a fixed place of work, the following types of injuries occurring on the premises are compensable:

(a) injuries while going to and from work;
(b) injuries before or after working hours;
(c) injuries at lunchtime.

If the injury occurs off of the premises, it is not compensable.

Injuries sustained while traveling between home and a fixed site of employment are generally not compensable. This general rule is coming under attack, and exceptions are being construed liberally. The rule does not apply to employees who do not have fixed places of employment. Exceptions to this "going and coming" rule include:

- "If the employer provides the transportation, then the injuries sustained by the employee are usually considered to arise out of employment. This exception is by no means universally applied, however. The Arizona appeals court decision in *Rencehausen v. Western Greyhound Lines,* denying compensation under these circumstances, represents the view of the minority.

- "If the employee has some duty at home or en route, such that the activity of traveling between work and home becomes an incident of employment.

- "If the employee is paid for his or her time or expense in traveling back and forth, then traveling is sometimes considered to be in the course of the employment. Some states have definitely rejected this exception to the going and coming rule, e.g., *Orsinie v. Torrance* (Conn.). The North Carolina Supreme Court in *Hunt v. State* apparently rejected the exception, but an appellate decision from that state, *Warren v. Wilmington,* held that the fact that an injured worker was being paid for her travel time was at least a factor in allowing her claim. The shift in attitude of North Carolina's judiciary reflects a general acceptance of this exception across the nation.

- "If the employee uses his or her own vehicle in the performance of job duties, then some courts have held that injuries sustained in that vehicle on the way to or from work are compensable. A California appellate decision, *Rhodes v. Workers' Compensation Appeals Board,* explains the rationale for this exception—which appears to be uniformly applied in similar cases: the employees are *required* by their employment to submit to the hazards of private motor travel, which otherwise they would have the option of avoiding.

- "If employees are required to take unusual hazards in traveling to or from work; for example, [injuries sustained while] crossing railroad tracks...are sometimes said to arise out of employment. This is not a well-recognized exception to the general rule of noncompensability of going and coming injuries and was definitely rejected by the Michigan Supreme Court in *Guastelo v. Michigan C.R. Co.* When [this exception is] invoked by a court to allow a claim, [it] is often limited to claims for injuries sustained in those areas directly under control of the employer (e.g., *Utah Apex Mining Co. v. Industrial Communication of Utah* (Utah))." [14]

Risk Factors

From the employer's point of view, all risks that can cause injury may be put into three categories: those associated with the employment, those personal to the employee or injured worker, and those which are neutral.

The group of those injuries connected with the employment are generally those more obvious types of injuries, such as the kinds of things that can typically go awry in modern factories, mills, mines, and construction; machinery breakdowns, explosives, extremities getting caught in machinery, equipment, or objects falling, and so on. Other things included in this group are more disease-related occupational injuries, such as black lung disease from mining, and chemical poisoning from pesticides and other sources.

In the personal risk category are those types of jobs or injuries that are so personal that, even if they happen while the employee is on the job, there is little possibility that the injuries could have been caused by the employment. An example of this is if an employee has a disease that is so advanced that she would have died whether or not she were at her employment, just because of the nature of her disease. Another example is if the employee for some reason has a contract for his death on him. If he were killed at or during work, this would obviously not have anything to do with work. And if an employee had terminal cancer and were to die at work instead of at home, this would be disallowed because he would have expired anyway.

The neutral-type risks are those that fall in between these two. Examples are an employee shot by a stray bullet at work, or struck by lightning, or bitten by a dog. Or a manager who is found face-down in a pile of accounting records and isn't that same healthy pink color you've come to know. His wife may have a compensable claim because her husband died at work of no apparent cause.

In a number of board decisions, awards have been made when the injury occurred because the job required the employee to be in what turned out to be a place of danger. Unexplained falls and deaths occurring in the course of employment have been awarded settlements and have been shown to be the responsibility of the employer, just because the injury or death occurred in the course of employment and arose out of the employment, without any evidence to the contrary.

Death Claims

Death benefits need to be related to the work environment. It needs to be established that the death was caused by a work-related injury or occupational disease. The death benefit obviously does not go to the worker, but instead goes to the family of the worker. It is provided to help them financially recover from the loss.

The benefit is not one that will totally replace the income that was generated by the injured worker. It is more of a stopgap type of payment.

Summary

There are a number of things to look at when thinking about injuries. These are areas that need to be addressed by all management personnel. You need to be aware of those areas that might not seem like work-related injuries, but in reality are.

Types of Injuries

Whether you are the employee or the manager, you are aware of the myriad varieties of injuries which can occur in the workplace.

Sudden impact injuries happen in an instant at a specific place and time—but usually with no warning.

Cumulative or repetitive injuries are labeled under multiple names, but the course of the injury is usually the same.

The last category of injuries is, in reality, a disease—namely, occupational disease. Occupational diseases may develop after long-term, repeated exposure to chemicals such as pesticides, asbestos, or lead. Work-induced heart attacks attributable to workplace stress are also included as an occupational disease.

With proper training and attention to safety, your work force can reduce the number of sudden impact injuries. Even in the best of organizations, an accident may occur. Experience has shown that workplace injuries increase when safety is a low priority. Programs that feature one to three days of training on correct equipment operation, safety recognition, and the right way to lift loads have proven effective in reducing injuries and the amount of workdays lost, and also in holding down workers' compensation premiums. A pro-active stance of teaching safety compliance as being in the best interest of the employee has the highest success rate. You follow the traffic signals by stopping at the red light because you know it is the right thing to do. However, following the rules because of fear of reprisals—such as receiving a ticket from the local police office—leads to a feeling of powerlessness in the face of a merciless authority. This is also true with safety compliance: pro-active programs have the highest success rate with employees and employers.

Whether you're doing a job that requires heavy lifting, prolonged sitting, or something in between, you personally will benefit by being injured less frequently than your fellow employees if you keep an eye on preventing accidents.

As stated earlier, cumulative or repetitive injuries have many different general names, such as:

- repetitive strain injuries
- cumulative trauma injuries
- overuse syndrome
- occupational cervicobrachial disorders

Repetitive-type injuries are a group of painful disorders of the muscles, tendons, and nerves. Names like carpal tunnel syndrome, tendonitis, thoracic outlet syndrome, and tension neck syndrome are always included in this category.

Since most, if not all, of the work needs the use of the arms and hands, a greater percentage of repetitive injuries appear in the hands, wrists, elbows, neck, and shoulders than in any other anatomical area. There are also cumulative trauma injuries in the back, legs, hips, ankles, and feet that result from repetitive injuries.

Are these types of injuries common? "They are considered to be the leading cause of significant human suffering, loss of productivity, and economic burdens on society. In 1987, over 20,000 workers in Ontario, Canada received compensation for new cases of repetitive motion injuries, accounting for 600,000 days of lost work. In 1982, a survey by Simon Fraser University and the United Food and Commercial Workers found that over 30% of the cashiers throughout British Columbia had suffered from an injury of this nature." [15]

Another recent study conducted by the National Center for Health Statistics shows that one in five workers suffered a week or more of back pain during the year, and 22 percent had trouble with their hands, including carpal tunnel syndrome.

These figures were part of a 1988 survey of 44,233 adults. The reports were compiled during the summer of 1993, and some of the results are listed below.

- One of every 15 workers was injured on the job during the year.
- Back injuries were the most common, with 18 percent of the surveyed workers reporting injuries.
- 22 percent of the injuries were to the hand, wrist, and finger.
- Sprains and strains were the most common injuries, with 28 percent reporting them.

- 21 percent reported cuts or punctures.
- One in 11 had to change jobs or job tasks as a result of injuries.
- 22 percent of men and 10 percent of women worked at jobs that required more than four hours of strenuous physical activity each day.
- 40 percent of male and 36 percent of female workers reported that they had to bend or twist their hands for at least four hours each day.

The risk factors of repetitive motion injuries generally arise from ordinary hand and arm movements such as bending, straightening, gripping, holding, clenching, and reaching. These movements unto themselves are not really a problem; the problem arises when these motions are done in a repetitive manner like that which can occur in the work environment. It is even worse because most of the time in the workplace, these motions are done in a manner requiring a great deal of force and continuous repetition, with little time to recover between the movements.

When you look at the average factory worker, construction worker, or machinist, you will find that all these types of occupations require repetitive, forceful motions. These fields of endeavor rank high on the list of types of work that can cause a repetitive motion injury.

When viewing activities in the workplace that are associated with repetitive motion injuries, you must include the following:

- Fixed or constrained body positions;
- Continual repetition of movements;
- Force concentrated on small parts of the body, like the hand or wrist; and
- A pace of work that does not allow sufficient recovery between movements.

The above-mentioned activities on their own are unlikely to cause a repetitive motion injury. It is the combination of these types of activities that can cause the problem. Heat, cold, and vibration also are contributors to this type of injury. It is up to you, as the employer, manager, or supervisor, to do everything possible to eliminate the chances of these types of injuries happening.

You need to look at the work environment. Determine for yourself, or get an outside evaluation, as to whether the types of jobs being done in your workplace are predisposing your employees to Repetitive Motion Injuries (RMI), and thus setting you up for a workers' compensation claim.

As you should be well oriented with the types of jobs your employees are doing, you should inform them of when they need to rotate the type of work they're doing; if they should take more frequent, short time breaks; and to let you know if their work stations should be redone to help prevent injuries.

Having the employees wear specific job-related preventative equipment is also a way to prevent injuries. It is important to note here that even though most of the injuries are structural, chemical-based repetitive injuries also happen very frequently. These employees will need special types of equipment and special work regimens to reduce the possibility of work injuries. Secretaries and data entry personnel need preventative equipment too, as they commonly get repetitive injuries—namely, carpal tunnel syndrome.

When looking further at RMI, you will see that body position comes into play. The position of the body when it is performing a particular job must be scrutinized. Tasks that require the joints involved to be at the extreme range of their motions greatly increase the chances of an RMI-type injury occurring. (See table 4.1 (below) and the corresponding figures (on page 49) for illustrations of this.)

Table 4.1

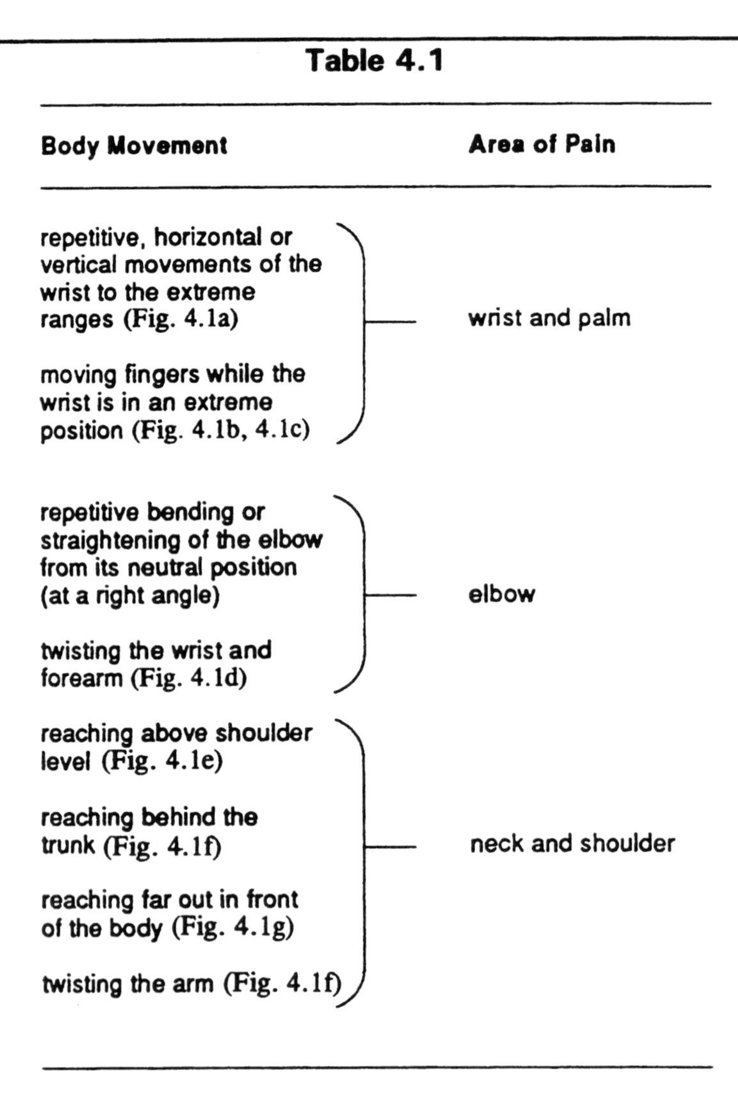

Body Movement	Area of Pain
repetitive, horizontal or vertical movements of the wrist to the extreme ranges (Fig. 4.1a) moving fingers while the wrist is in an extreme position (Fig. 4.1b, 4.1c)	wrist and palm
repetitive bending or straightening of the elbow from its neutral position (at a right angle) twisting the wrist and forearm (Fig. 4.1d)	elbow
reaching above shoulder level (Fig. 4.1e) reaching behind the trunk (Fig. 4.1f) reaching far out in front of the body (Fig. 4.1g) twisting the arm (Fig. 4.1f)	neck and shoulder

Source: Canadian Centre for Occupational Health and Safety, *Nature of Injury by Occupation.* (Toronto: Canadian Centre for Occupational Health and Safety, 1985).

Hazardous Repetitive Movements

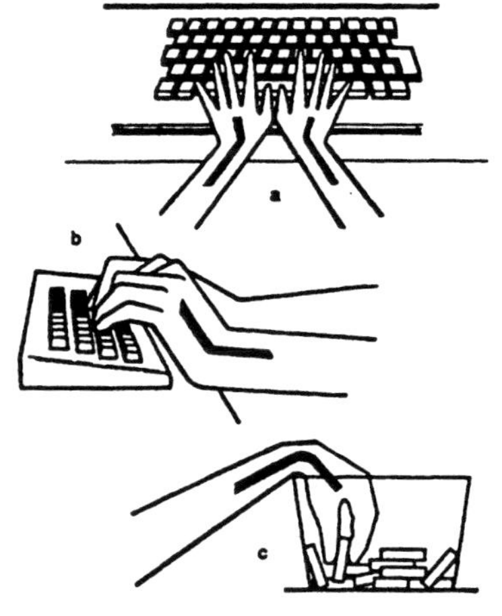

Hazardous movements of the hand

Figures 4.1a - 4.1c

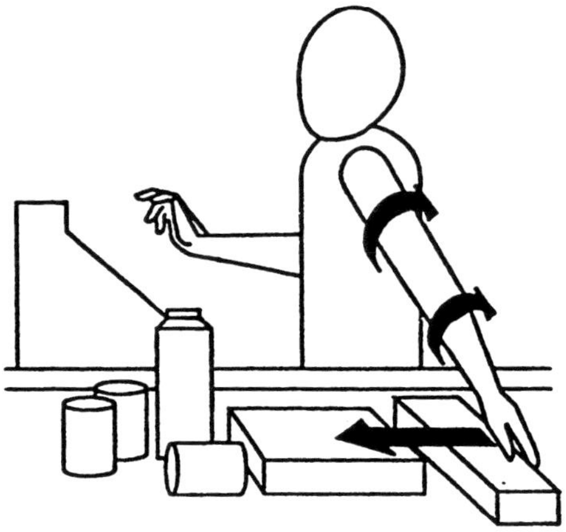

Reaching behind the trunk

Figure 4.1f

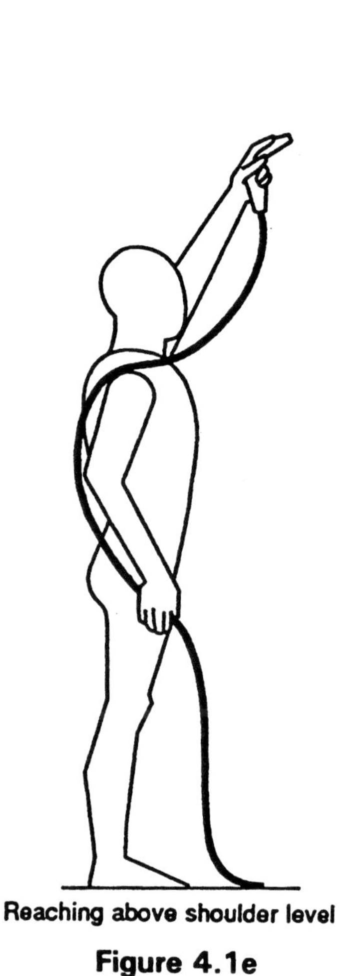

Reaching above shoulder level

Figure 4.1e

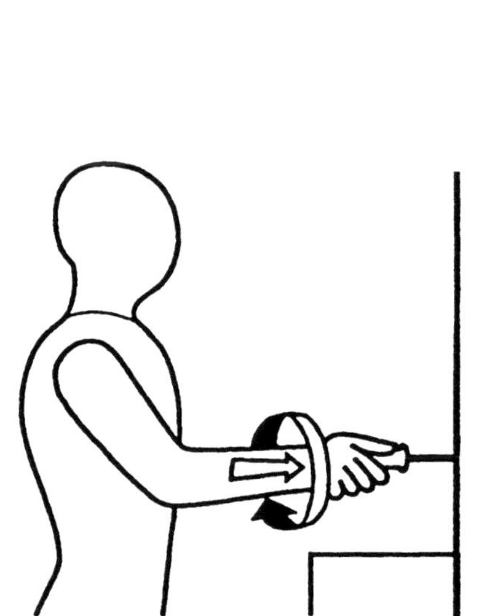

Exerting force while extending forearm

Figure 4.1d

Reaching forward

Figure 4.1g

Figures 4.1a - 4.1g (corresponding with table 4.1)

Source: Canadian Centre for Occupational Health and Safety, *Nature of Injury by Occupation*. (Toronto: Canadian Centre for Occupational Health and Safety, 1985).

Another aspect of body position that can contribute to RMI occurs when the neck and shoulder are in a relatively fixed position. While performing any controlled activity using an upper limb, the worker must stabilize the shoulder-neck region. After doing the same kind of motion over an extended period of time, the muscles of the neck and shoulder girdle are forced to contract, and they stay contracted in order for the employee to continue to hold his current body position. The contracted muscles then squeeze the blood vessels, which in turn restrict the blood all the way down to the working muscles of the hands, where, because of the intense muscular effort, the blood is needed the most. This process continues until the particular task or motion is completed. The result is twofold: (1) the neck and/or shoulder muscles become fatigued even though they are not moving, and (2) the lack of blood supply to the hands contributes to increased pain in the neck area. Also, while all this is going on, the reduction in blood supply to the rest of the upper extremities increases the fatigue in all the muscles, making the areas more vulnerable to injury.

As stated previously, those employees who are doing jobs requiring a great deal of repetitive motion are at the greatest risk of having a RMI. Understand that while repetition of motion is not the only factor in RMI, it is the most likely one, and thus it is the single highest risk factor. But when you have an employee doing repetitive movements in his normal job, other risk factors of RMI are always involved as well, such as fixed body position and force: the worker has to maintain the shoulders and neck in a fixed position and exert some force in order to perform the job.

It needs to be stressed here that work involving the same motion over and over is very tiring, both to the individual as well as to the various structures of the body. Not only is this repetition physically tiring, but it can also be mentally tiring. This is when the employee goes into "automatic". This mental attitude leaves the employee wide open for injury, either a repetitive type or the sudden type of injury.

The other problem with this type of repetitive work is that the worker's body has little or no time to recover from one task before the same task is upon him again. When doing this type of work, even if it is minimal labor, the challenge to the body increases steadily; thus, fatigue sets in more quickly, and the chances for injury increase as well.

The occurrence of these injuries is not sudden, but happens gradually over an extended period of time. The muscles and ligaments are gradually stretched, strained, over-exerted, and fatigued to the point where they finally show all the symptoms of a RMI. Inflammation of the area sets in, causing an increase in the pressure to that area, and thus again increasing the probability of injury.

The three basic types of injuries occur to:

- muscles
- ligaments
- nerves.

Injuries to the muscles have their greatest occurrence from the repeated contraction of the muscles. This contraction causes an increase in the usage of sugars and produces by-products like lactic acid, which are usually easily removed from the blood. When there is an extended muscle contraction, however, the blood flow is restricted, and the by-products are not as easily, if at all, removed from the muscle. This gradual build-up of substances causes irritation and pain—how much pain and irritation depends directly on the duration of muscle contractions and the amount of time between activities. The shorter the amount of recovery time, the greater the amount of pain and irritation; thus, people doing highly repetitive type work have a higher risk of this type of injury.

Another area that repetitive motion can affect is the tendons. Anatomically, tendons are bundles of fibers that connect muscles to bones. Problems of the tendons are caused not only by repetitive motions, but also by awkward postures. Tendons are grouped into two major categories: those with sheaths, found mostly in the hand and wrist; and those without sheaths, generally found in the shoulder, elbow, and forearm. (See figure 4.2 below.)

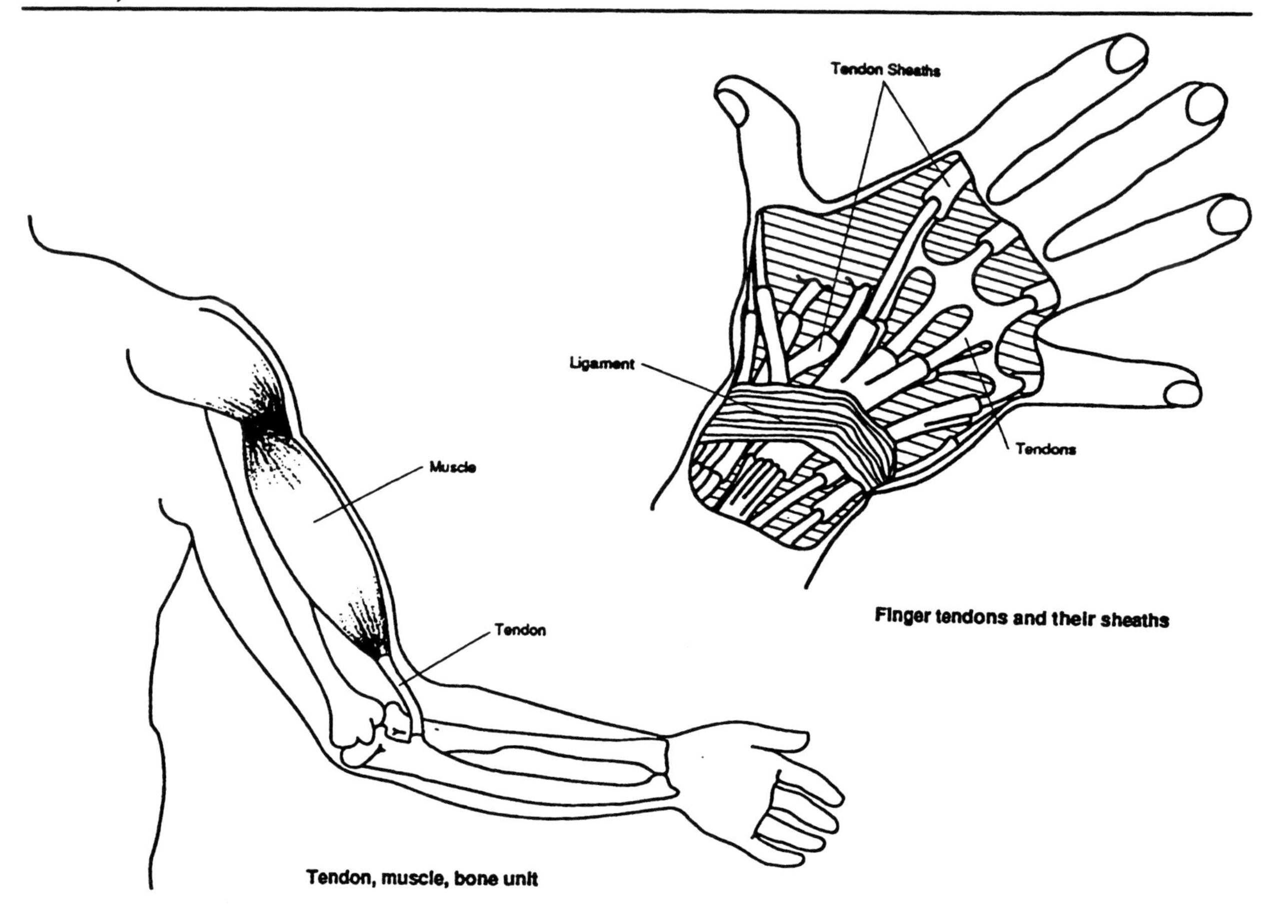

Figure 4.2

Source: Canadian Centre for Occupational Health and Safety, *Nature of Injury by Occupation*. (Toronto: Canadian Centre for Occupational Health and Safety, 1985).

The tendons of the hand are encased in sheaths into which the tendons slide. The inner wall of these sheaths produces a lubricating substance that helps the tendon slide. In work that requires excessive movement, this system of lubrication may not work properly. Failure of this system to work causes an increase in friction in the passage of the tendon; and, just like with any moving object that needs lubrication, the risk of failure of the particular part is inevitable when friction builds up and motion continues. Inflammation of the tendon sheath is called tenosynovitis. When inflamed, the tendon sheath may swell up with the thickened lubricating fluid and cause a bump under the skin. This is referred to as a ganglion cyst.

The tendons that are without sheaths are also open to problems of repetitive motion and awkward postures. When a tendon is repeatedly tensed, it is not uncommon for some of its fibers to be torn apart. At this point the tendon becomes thick and bumpy, causing inflammation. This condition is called tendonitis.

Tendonitis is a general term describing inflammation of a tendon. In some cases, such as in the shoulder, tendons pass through a narrow space between bones. A sac, called a bursa, filled with lubricating fluid is between the tendons and the bones to serve as an anti-friction device. As tendons become increasingly thick and bumpy, the bursa is subject to a lot of friction and becomes inflamed. Inflammation of a bursa is called bursitis.

Repetitive motion can also cause harm to the nerves. Nerves carry signals from the brain to control the activities of muscles. They also carry information about temperature, pain, and touch from the body to the brain, and they control bodily functions such as sweating. Nerves are surrounded by muscles, ligaments, and tendons. With repetitive motions and awkward postures, the tissues surrounding nerves become swollen and squeeze or compress the nerves. Compression of a nerve causes muscle weakness, sensations like "pins and needles", and numbness.

Symptomatically, repetitive motion injuries progress into a number of different phases or stages. Of course, pain is usually the first and most common symptom. In some cases, symptoms such as redness and swelling of the injured area will occur. Other symptoms of note are "pins and needles," numbness, change in skin color, and a decrease in the sweating of hands (when the hands are affected).

The following is a brief description of the different stages that these types of injuries may progress through.

Early Stage: Aching and tiredness of the affected limb occur during the work shift but disappear at night and during off-work hours. No reduction of work performance.

Intermediate Stage: Aching and tiredness occur early in the work shift and persist at night. Reduced capacity for repetitive work.

Late Stage: Aching, fatigue, and weakness persist at rest. Inability to sleep and to perform light duties.

Realize that these symptoms and patterns are generalities, and that not everyone will go through these in the exact same way or time. Because of the vagueness of the symptoms, it will be difficult to truly determine what stage someone is in. The first pain experienced is a signal that the muscles and tendons should rest and recover. Otherwise, an injury can become chronic in nature, and sometimes irreversible. The earlier the symptoms and problems are recognized, the better the chances of a full recovery.

Table 4.2 (on page 54) gives a good, brief description of disorders and syndromes, what the basic symptom patterns are like, and what types of jobs can cause them.

In order to properly evaluate these types of injuries, it is necessary to do a number of things. First, conduct an evaluation of the workplace, including a job description and some form of visual inspection, to see what the job looks like while it is being performed. Medical intervention should come next. This can be in the form of a simple examination, or it can be as complex as laboratory and electrical testing to see whether there has been any nerve or muscle damage. Different tests that are used are EMG and SSEP, as well as the occasional MRI. (These tests will be explained in greater detail in the chapter on medical care.)

Treatment for these injuries can range from very conservative to surgery, depending on the extent of the damage and how soon the problem is diagnosed and treated. Conservatively, the initial form of treatment can be as simple as just restricting the motion of the joint or area involved. This can be done by way of a brace or by changing the current job the employee is doing. Along with this is the inclusion of heat or ice therapy, again depending on the findings of the doctor involved.

It should be quite apparent that the easiest way to treat these types of injuries is to either find another job for the injured worker, or to change the way the job is being done. Look at the ergonomics of the situation. Modify them to help prevent further injuries to other employees and to reduce workers' compensation costs.

Repetitive motion injuries are the fastest growing form of industrial injury, and most of them can be prevented if the company is willing to spend a few extra dollars and look at the situation instead of blindly doing things as they have always been done.

Carpal Tunnel Syndrome

The only syndrome or disease that will be discussed in detail in this book is carpal tunnel syndrome. The reason for detailing this syndrome is that it is on the upswing, as far as work-related injuries are concerned. As employers and managers, it is very important

Table 4.2: Identified Disorders, Occupational Risk Factors, and Symptoms

Disorders	Occupational Risk Factors	Symptoms
Tendonitis/Tenosynovitis	Repetitive wrist motions Repetitive shoulder motions Sustained hyperextension of arms Prolonged load on shoulders	Pain, weakness, swelling, burning sensation or dull ache over affected areas
Epicondylitis (Elbow Tendonitis)	Repeated or forceful rotation of the forearm and bending of the wrist at the same time	Same symptons as tendonitis
Carpal Tunnel Syndrome	Repetitive wrist motions	Pain, numbness, tingling, burning sensations, wasting of muscles at base of thumb, dry palm
DeQuervain's Disease	Repetitive hand twisting and forceful gripping	Pain at the base of the thumb
Thoracic Outlet Syndrome	Prolonged shoulder flexion Extending arms above height Carrying loads on the shoulder	Pain, numbnes, swelling of the hands
Tension Neck Syndrome	Prolonged restricted posture	Pain

Source: Canadian Centre for Occupational Health and Safety, *Nature of Injury by Occupation*. (Toronto: Canadian Centre for Occupational Health and Safety, 1985).

that you know about how it can be both prevented and treated. The syndrome comes about from all types of work that are repetitive in nature—from factory work, secretarial work, or other types of job duties.

It is a condition that affects the hands and the wrists. The carpal tunnel is a space in the wrist where tendons and the median nerve pass. These go into the hand for the purpose of controlling the fingers and the thumb. The whole area is surrounded by the wrist bones and a rigid ligament that holds the bones together.

The tendons attach the muscles to the bones of the hand, allowing for the motion of the fingers by way of the muscles to the bones. The median nerve carries the signals from the brain to allow for this motion. It also carries information about pain, temperature, and touch from the hand to the brain. Control of the sweating mechanism of the hand is also handled by the median nerve. The areas of the hand that are innervated by the median nerve are the thumb, index, middle, and ring fingers.

In the tunnel, the tendons of the fingers surround the median nerve. If an injury or irritation to this area occurs, either at work or elsewhere, it will generally lead to inflammation of the area. This inflammation will put pressure on the median nerve, causing injury to it. The symptoms reported on injured median nerves range from pain to numbness to tingling, or even to clumsiness of the hand.

As stated earlier, the tendons are encased in sheaths, which have a method of self-lubrication. When the tendons are injured or irritated, the process of lubrication can be disrupted, causing an increase in friction in the area. The increased friction causes inflammation, and the inflammation then leads to the pressure on the nerve, which in turn leads to symptoms. If the above process is repeated, the continuous inflammation can lead to the body producing fibrous tissue. The fibrous tissue then leads to a thickening of the tendon sheath, which can dramatically reduce tendon movement. (See Figure 4.3 below.)

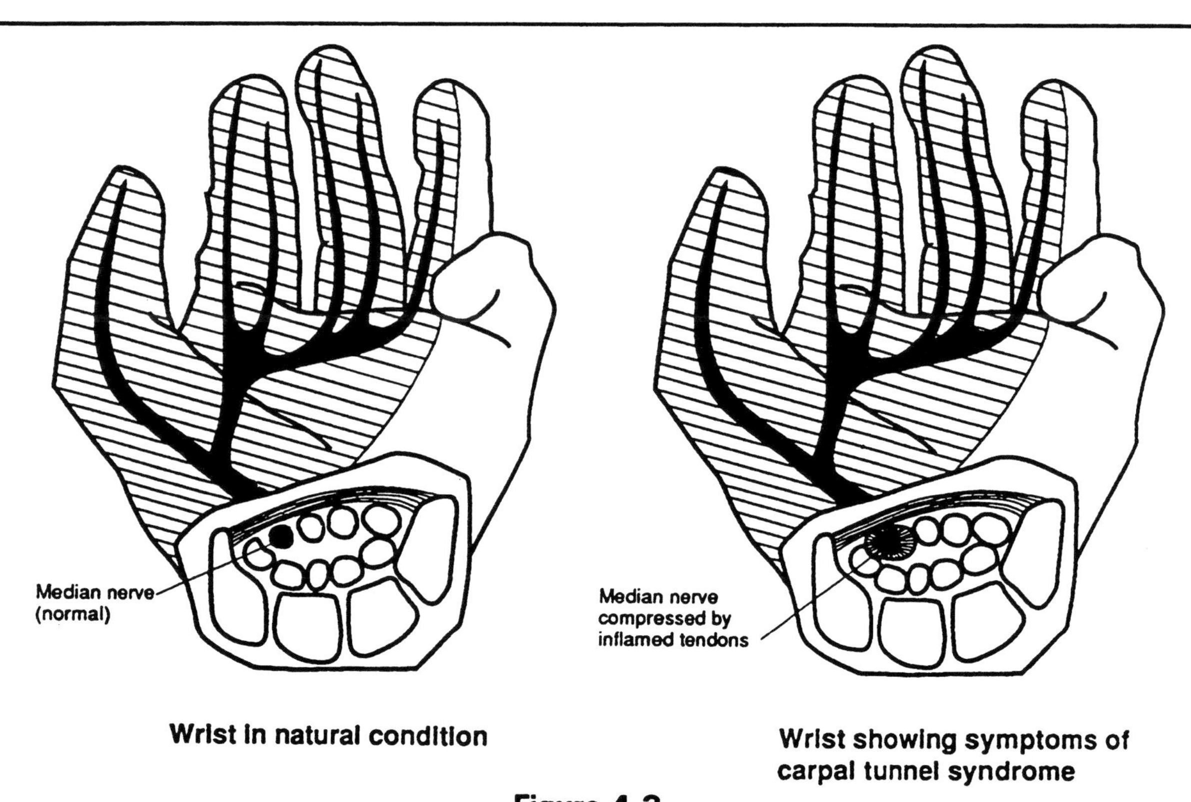

Figure 4.3

Source: Canadian Centre for Occupational Health and Safety, *Nature of Injury by Occupation*. (Toronto: Canadian Centre for Occupational Health and Safety, 1985).

"A survey of 982 supermarket checkers revealed that 614, or 62.5%, had symptoms of carpal tunnel syndrome. Of 788 meat handlers, 117 had surgical treatment for carpal tunnel over a twelve year period. A survey of 400 American hand surgeons reported that each surgeon performed an average of 65 operations for carpal tunnel syndrome each year." [16] Recent studies have shown that carpal tunnel syndrome doesn't just happen to laborers. It is quite common in people who are doing data entry or heavy secretarial work on a daily basis, among others.

Different types of musculoskeletal injuries in any workplace, from single incident traumas to cumulative or repetitive traumas, are not only inevitable, but they are also unstoppable. Mistakes will happen and employees will get careless, no matter how much training, prevention, or ergonomics they have been taught. Your responsibility as the manager or supervisor is not only to try and keep the volume of injuries down, but also, once the injuries have happened, to keep them contained and well managed.

Injuries are going to happen in any business under any circumstances, and the best that any of us can do, whether in management, ownership, employment, or from the medical end, is to try to contain them and keep their numbers to a minimum. Be aware of the symptoms that employees may discuss. Have ergonomic studies done on your premises, so that you can help avoid the simple injuries. Let a medical provider have access to your workplace so that he or she can help you look at the biomechanics of the jobs being done. These simple acts can save you not only directly, but also indirectly, in terms of lost time of employees, lost production, and other aspects of your business.

Diseases

Besides the basic type of injury, which is musculoskeletal or cumulative in nature, there are a great variety of other diseases that are also compensable within the typical state's workers' compensation system.

"Certain types of diseases are very specific or peculiar to an industrial trade or process. To the employee, this means that the worker is under a greater risk of exposure to specific diseases than someone in the general public. The Ohio Supreme Court addressed this definition in *Ohio Bell Telephone Co. v. Krise,* which allowed a claim for histoplasmosis incurred by a telephone lineman's exposure to pigeon droppings. The case contains a discussion of the phrase 'peculiar to an industrial trade or process'.

"[The] following are a list and some observations on coverage of some specific ailments or conditions.

- "Hernias are often compensable and, in those jurisdictions that differentiate injuries from occupational diseases, they are usually classified as injuries. Like many states, Florida's law provides specific limitations on length of compensation for hernia victims *(Plant City Steel v. Grace)*, and on conditions for coverage of this condition *(Metropolitan Dade County Bd. of Commissioner v. Wyche)*. Mississippi's provision is similar *(Bechtel Constr. Co. v. Barlett).*

- "Hearing loss, allegedly from noisy environments, is usually so gradual and its etiology so difficult to ascertain that several states have set out special restrictions on compensability.

  "A majority of states treat hearing loss as an occupational disease rather than an injury, but the Pennsylvania Supreme Court in *Hinkle v. H.J. Heinz Co.* treated it as an injury on 'repeated trauma' basis. Some states have set up separate statutory provisions for hearing loss, the most impressive of which is New York's, which provides a separate waiting period, definition of disablement, apportionment of liability among employers, and minimum-exposure guideline. The provision in Wisconsin's 'Occupational Deafness' statute, that no compensation is payable for temporary total or temporary partial disability resulting from occupational deafness, is not unusual. Ohio provides for no permanent-partial disability awards for less than total loss of hearing in one ear.

- "Allergies are subject to varying treatment among the states according to particular circumstances; however, some principles seem fairly well established. The mere fact that an individual has an allergy before beginning employment does not render subsequent manifestations of the allergic condition noncompensable. Generally speaking, if work occasions an exposure to an irritant to which the worker is already allergic, that specific incident is a compensable injury but the underlying condition is not. On the other hand, an individual who develops a sensitivity to an irritant because of work exposure will probably have a compensable occupational disease claim in most U.S. jurisdictions. Moreover, it appears that allergies are compensable if employment exposes a worker to an irritant to which he is unknowingly allergic. In *Webb v. New Mexico Publishing Co.* (N.M.), such an allergy claim was allowed but as an injury, not as an occupational disease.

  "Many states provide in their statutes for 'contact dermatitis', a condition resulting from exposure to chemical irritants. Two states have rejected allergic reactions as occupational diseases: [Maryland, in the case of *Kelly-Springfield Tire Co. v. Roland,* and Missouri, in *Sanford v. Valier-Spies Milling Co.*].

- "Mental disabilities or stress-caused illnesses present a special problem. As a general rule, psychiatric conditions that are proved to be the result of an already allowed injury or occupational disease are compensable as part of the underlying claim. The psychiatric condition is allowed as a 'flow-through' disability, much like traumatic arthritis is commonly allowed as a 'flow-through' injury in claims initially allowed for fractures. Although special procedures—such as the claimant's completing and filing of an affidavit regarding psychiatric history—may sometimes be required, there is no legal problem with allowance of such conditions as part of an already allowed claim." [17]

A legal problem sometimes arises when mental stress or emotional trauma is the only cause of the disability. In states which still require some sort of physical trauma or sudden occurrence in injury claims, emotional trauma obviously cannot form such a predicate, since they are, by definition, not physically traumatic. In some jurisdictions, claimants must attempt to prove that their emotionally-induced disabilities are occupational diseases. "They therefore must establish that their employment was of such a nature as to expose them in a 'peculiar' way to such emotional distress. Not surprisingly, allowances of claims for mental distress as occupational diseases have not fared well for claimants. *Voss v. Prudential Ins. Co.* (N.J.) and *Szymanski v. Halle's* (Ohio) have refused to recognize mental-distress disabilities as injuries, while *Transportation Ins. Co. v. Maksyn* (Tex) refused to recognize such disability as an occupational disease.

"However, in those jurisdictions that provide no legal impediment to claims for mental distress, such claims have often been allowed as injuries when the work-related incidents giving rise to the disability were particularly distressful.

"Some authorities have noted a trend toward allowance of mental-distress claims, whether as an injury or as an occupational disease: [for example], *Carter v. General Motors Corp.* (Mich.)." [17]

Cumulative Injury Claim Statistics

The president of a large company has a heart attack on the golf course and then files a disability claim for workers' compensation, saying that the cause of her heart attack was based on 25 years of occupational stress. A firefighter files a similar claim for his case of high blood pressure resulting from his very tough and demanding occupation. These types of claims, those loosely based on job-related stress, are beginning to be more and more common and pose a looming threat to an already overburdened workers' compensation system.

The workers' compensation system was created to handle structural/anatomical injuries such as broken bones, sprains/strains, and other single-incident injuries. Now, the system is evolving to include repetitive motion injuries and long-term, stress-related diseases such as hypertension, heart attacks, etc.

The big problem is distinguishing between work-related and age-related injuries. Was the employee's heart attack work-induced or was it because she was seventy-five? Would your answer be different if the employee was thirty-five? Should the employer be on the hook for all heart attacks by employees? Good question.

When you start looking at the different types of cumulative injury claims, you will find that the list of causes goes as follows: skin/contents, lower back problems, internal organs, psychiatric, and soft neck tissue. (See figure 4.4 on page 59.)

Five Most Frequent Injuries
Cumulative Injuries

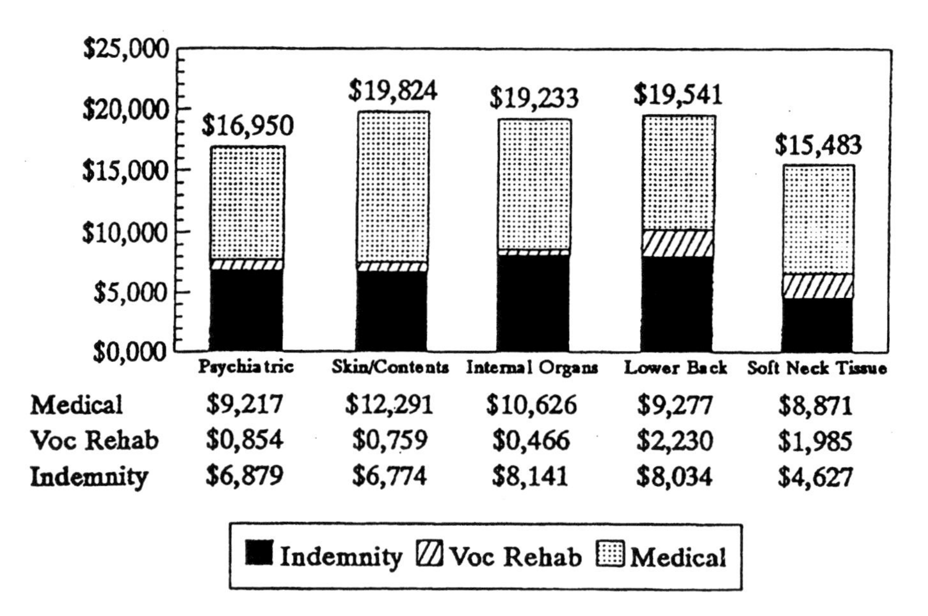

	Psychiatric	Skin/Contents	Internal Organs	Lower Back	Soft Neck Tissue
Medical	$9,217	$12,291	$10,626	$9,277	$8,871
Voc Rehab	$0,854	$0,759	$0,466	$2,230	$1,985
Indemnity	$6,879	$6,774	$8,141	$8,034	$4,627

Figure 4.4: Incurred Benefit Costs

Source: M.L. Miller, *Cumulative Injuries and Specific Injuries in California, A Report to Industry.* (San Francisco: California Workers' Compensation Institute, 1995).

A study done recently by the California Worker's Compensation Institute, based on cumulative versus specific injuries in 1990, shows some distinctions between these two types of injuries. (See figure 4.5 on page 60). The study revealed the following:

1. The most common cumulative injury claim involved psychiatric injuries. As of 1994, the criteria for making a psychiatric claim were made dramatically tougher. In the past, a claim could be made if the worker could show only a 10 percent work involvement that caused him to make the stress or psychiatric claim. The new law changed that to the worker needing to show 51 percent work involvement. This change in the law has already proven to have reduced the number of these types of claims.

2. Attorneys were much more likely to be the first source of notice to the insurance companies in a cumulative injury claim than in a specific injury claim, creating an immediate relationship between the injured worker and the employer or insurer. This fact alone is a point the author has stressed throughout this text. When an adversarial relationship starts from the beginning, the nightmares of workers' compensation can quickly and easily escalate. If you are told by an attorney that the employee is making a claim, or if you are served with the appropriate paperwork, you need to contact the employee and try to resolve the situation as amicably as possible. This will not only save you money, but it can also save you many headaches.

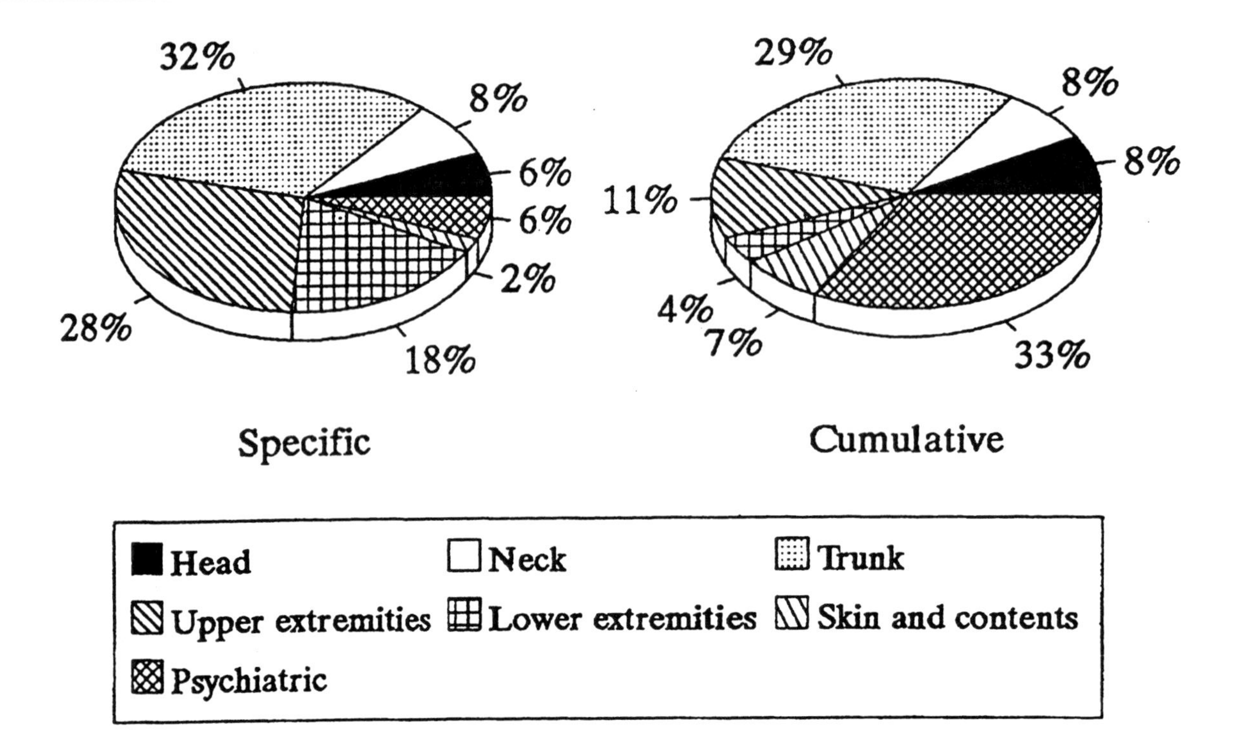

Figure 4.5: Types of Specific and Cumulative Injuries

Source: M.L. Miller, *Cumulative Injuries and Specific Injuries in California, A Report to Industry.* (San Francisco: California Workers' Compensation Institute, 1995).

3. Insurers were more likely to initially respond to a cumulative injury claim with a delay or denial, which may cause or intensify the worker's perceptions of unfairness, feelings of anger, and the perceived need for litigation.

Neither you nor the insurance company want to be on the hook for injuries that may not be your fault. Neither of you want the possible hostility that can grow out of a contested claim, yet both of you need to be protected. Again, the answer may be to keep in touch with your employees—let them know how you feel and show them the concern that you would want to receive. Also, it is important for you to remember that you control the insurance company and not the other way around. If you want them to pay a claim, tell them to pay it. They have to.

4. Employees claiming cumulative injuries were more likely to retain a self-procured treating physician at the time of injury without approval of the employer or the insurer, and insurers were more likely to object, creating a crucial point of dispute.

When an employee feels that you are not going to do anything about his claim, he is going to get medical or chiropractic care with or without your approval. This, once again, creates negative feelings between you and the employee, whether they are justified or not.

5. Litigation was nearly universal with cumulative injury claims, and occurred much earlier than with specific injury claims.

6. The proportion of medical-legal evaluations and diagnostic tests which went unpaid (at least initially) was twice as high for cumulative injury cases as for specific injury claims.

When litigation enters the scene, costs can very easily skyrocket. Both sides are trying to prove their point; thus medical experts get into the scene and generally do their own tests with the facilities they prefer. Each expert also writes reports which have his or her own costs (see figure 4.6 below). All of this escalates the costs to every employee, employer, and person within a given state, because these increased costs get passed on from the insurance company to the employer and so on down the line.

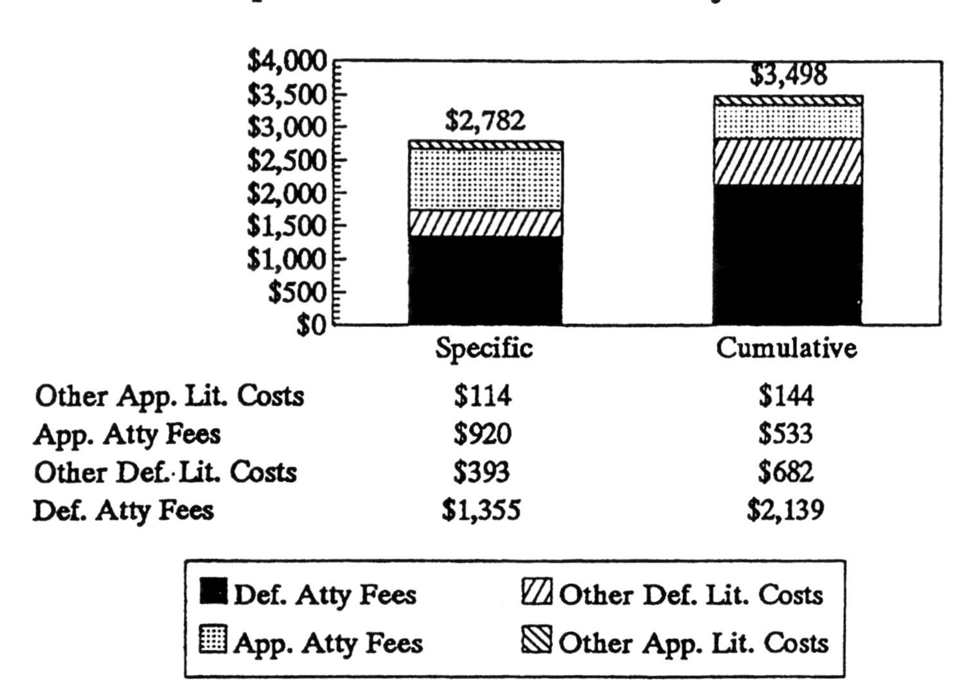

	Specific	Cumulative
Other App. Lit. Costs	$114	$144
App. Atty Fees	$920	$533
Other Def. Lit. Costs	$393	$682
Def. Atty Fees	$1,355	$2,139

Figure 4.6: Incurred Litigation Costs

Source: M.L. Miller, *Cumulative Injuries and Specific Injuries in California, A Report to Industry.* (San Francisco: California Workers' Compensation Institute, 1995).

It was also shown, according to the Workers' Compensation Insurance Rating Bureau (the California Insurance Commissioner's designated statistical agency), that between 1983 and 1991, the incidents of cumulative injuries and occupational diseases per 1000 workers in California more than tripled. (See figures 4.7 and 4.8 on page 62.)

Specific and Cumulative Injuries
Closed claims only

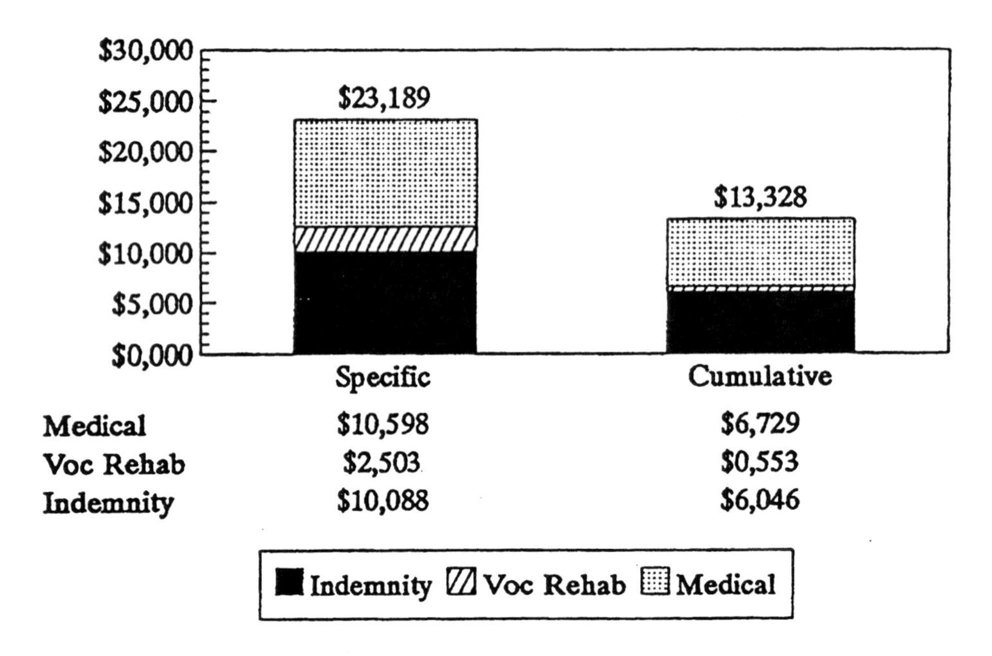

Figure 4.7: Incurred Benefit Costs

Source: M.L. Miller, *Cumulative Injuries and Specific Injuries in California, A Report to Industry.* (San Francisco: California Workers' Compensation Institute, 1995).

Specific and Cumulative Injuries
Accepted claims only

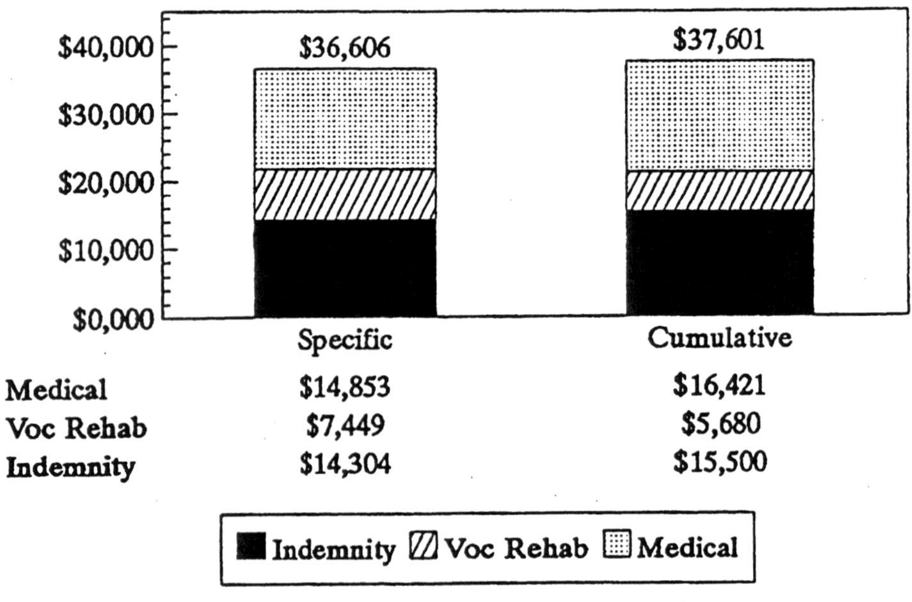

Figure 4.8: Incurred Benefit Costs

Source: M.L. Miller, *Cumulative Injuries and Specific Injuries in California, A Report to Industry.* (San Francisco: California Workers' Compensation Institute, 1995).

The following is a small test about cumulative trauma for your employees. It is designed to see if they are experiencing particular symptoms that could mean they have a cumulative trauma disorder.

1) Within the past month, have you had repeated feelings of numbness, tingling, or "pins and needles" sensations in one or both hands?

Yes ____ No ____

2) Within the past month, have you had repeated feelings of soreness or pain in either forearm or elbow?

Yes ____ No ____

3) Within the past month, have you had repeated feelings of pain, discomfort, burning, or tingling in your shoulders?

Left Shoulder Yes ___ No ___
Right Shoulder Yes ___ No ___

4) Have any of the above symptoms (numbness, tingling, soreness, or pain) caused you to awaken while sleeping?

Yes___ No___

5) What time does your discomfort regularly occur?

Morning Yes ___ No ___
Afternoon Yes ___ No ___
Evening Yes ___ No ___
Night Yes ___ No ___

6) Does discomfort in your wrist, arm, or shoulder interfere with your daily activities (eating, writing, sports, etc.)?

Yes ___ No ___

7) Have you ever received medical treatment for this pain and/or discomfort?

Yes ___ No ___

8) Have you ever received medical help (from either the company or a private doctor) for any of the following?

Carpal Tunnel Syndrome Yes ___ No ___
Ganglionic Cyst Yes ___ No ___
Tendonitis Yes ___ No ___

9) If yes to (8), have you ever had surgery for any of these conditions?

Yes ____ No ____

10) Does your present job require arm, hand, or finger actions to be repeated many times during each hour and work shift?

Yes ____ No ____

Summary

You not only must consider single action accidents in workers' compensation (for instance, a crane falling and hitting an employee), but you also need to be aware of repetitive motion injuries and occupational diseases. These can have the most devastating financial ramifications on any business. With your research and consulting experts' findings and suggestions, you can dramatically decrease the number and severity of repetitive motion injuries and decrease exposure to chemicals or situations which induce occupational diseases. Every positive, proactive action you take today can save you money and time in the future. Your goal is to provide an accident-free workplace where everyone can safely come to work, making money for one and all.

Chapter 5

Medical Care Choices

Now we can get into the arena of medical care. It is an area you *must* know about if you want to survive the workers' compensation system. Remember, this is the first area of the system that your employee sees after he or she has been injured.

Choice of medical care under workers' compensation is highly disputed and highly controversial. To begin with, you have traditional medical care—allopathic medicine, or osteopathic medicine. One of these is probably the doctor you normally see throughout your life for everything from a cold to a cut needing stitches to stomachache problems. These doctors are great when emergency care, drugs, or surgery is needed.

Beyond the allopathic physician, there are a number of different possibilities. The second most common choice of health care is chiropractic.

Chiropractic is concerned directly with the neuromusculoskeletal system. This is the nervous system, muscles, ligaments, and joints. Although many people are under the false impression that chiropractors only treat the spine, this is untrue. Chiropractors treat the spine and the other major joints of the body. The first two years of chiropractic college parallel the first two years of medical school. After that, chiropractors emphasize the neuromusculoskeletal system and how to adjust the joints. While chiropractors do receive training in surgery and pharmacology, most states do not license them to practice these skills. (This is not due to inadequate training, but instead to philosophical bias). Thus, chiropractors are thoroughly trained to deal with any and all sprain-strain type injuries.

Acupuncture, which in most states falls within the workers' compensation system, is another popular form of health care treatment. It is based on the meridian system. As defined by Taber's Cyclopedic Medical Dictionary, acupuncture is "a technique for treating certain painful conditions and for producing regional anesthesia by passing long thin needles through the skin to specific points." [18]

Other choices for medical care include homeopathy, Christian Science healing or prayer, and numerous others, depending on which state you are in.

Most companies choose an urgent care type facility as the primary place to send their injured workers. Depending on their preference for health care and the type of injury, this may or may not be the best place to send these workers.

Typically, an urgent care facility has either an M.D. or an osteopath as the primary physician. These are obviously the doctors best trained for emergency medicine, since they are trained to handle broken bones, bleeding conditions and injuries that require stitches, as well as to prescribe necessary drugs.

The real question here is whether all injuries should be considered emergencies. Obviously, a slip and fall, or a lifting injury after which the employee can still move but is sore, may not rate as demanding an emergency as a severe laceration. A sprain or strain under the above criterion might not be considered an emergency either, though it does require treatment just as surely. For injuries of this type, it might be advantageous to use a chiropractic physician. Chiropractors are trained primarily in knowledge of the musculoskeletal system. They are also trained in diagnosis, orthopedics, and neurology, as well as in many other topics. As the employer, you need to evaluate the type of injury that has been sustained before automatically sending the employee to one kind of doctor or another.

If your company's primary medical center is a chiropractic office, it is hoped that you would not send an employee with a broken arm to the chiropractor, but instead to an urgent care facility or hospital. In the same light, if your regular medical facility is an urgent care facility, it could be a great advantage to send the injured worker with a low back strain to the chiropractor.

One of the biggest concerns to business is how to control escalating medical costs in the workers' compensation environment. This issue affects every business and individual looking for adequate but cost-effective health care.

Over the last fifty years, there have been numerous cost containment studies comparing medical care to chiropractic care in the work environment. The following are a few of those studies that have been done in order to determine true cost-effectiveness between chiropractic and medicine. Other cost containment studies can be found in appendix A.

Cost Comparison Studies

In preparation of this text, the author contacted the American Medical Association regarding studies on the benefits of medical over chiropractic care. They stated that they had no such studies; they did not feel the need to do them, as they were the mainstream.

Studies of workmen's compensation records provide objective evidence of the efficacy of chiropractic health care in the treatment of industrial injuries. From data supplied by workers' compensation commissions, it has been dramatically demonstrated in comparisons of Doctor of Chiropractic (D.C.) and M.D. treatment of industrial injuries that those cases under chiropractic care showed reduced treatment costs, reduced compensation costs, reduced work-time losses, and reduced workers' disability and suffering.

Early Studies

"Two early comparisons of chiropractic vs. medical treatment for similar conditions were undertaken in Colorado and Florida. [See results summarized in the table below.]

State	Term of Study	Work time Lost: DC Treatment	Work time Lost: MD Treatment
Colorado	1948-49	2.6	4.9
Florida	1956	3.0	9.0

The Florida study utilized the largest number of cases to date (19,666), although the Colorado study was based on nearly 2,000 cases." [19]

Florida Study

"Based on Florida statistics, a study conducted in 1988 by the Foundation for Chiropractic Education and Research of Florida's closed workers' compensation claims for the 1985-86 fiscal year showed chiropractic [treatment] to be not only cost-effective, but quite efficacious as well. The study demonstrated that chiropractic case management, compared with standard medical case management, minimizes the impact of work-related back injuries and illnesses on prolonged absence from work and [also minimizes] excessive treatment costs." [20] (See table 5.1 on the following page.)

Table 5.1: Comparison of Non-Surgery and Surgery Claimants Treated by M.D.s and D.C.s

Group A Non-Surgery Claimants	Cases Treated by M.D.s	Cases Treated by D.C.s
Average days disabled per claimant	58	39
Average indemnity cost per claimant	$1,483	$1,159
Average total cost of care per claimant	$2,213	$1,204
Group A Plus Surgery Claimants	**Cases Treated by M.D.s**	**Cases Treated by D.C.s**
Average days disabled per claimant	59	39
Average indemnity cost per claimant	$1,534	$1,159
Average total cost of care per claimant	$2,352	$1,204

Source: American Chiropractic Association, *Chiropractic State of the Art.* (Washington, D.C., 1991-1992).

From the above studies and numerous others that have been done, it seems quite apparent that chiropractic may be a very good choice for the care of your injured workers. This is especially true when you consider the number of injuries that are sprains and strains.

As of 1991, nearly 45 percent of all work injuries were sprains and strains, with the back being the number one site of injury. In recognition of these statistics and of the fact that chiropractors specialize in helping these types of injuries, chiropractic should be the first choice for these cases. It should also be noted here that 75 percent of all "work comp" injuries are neuromusculoskeletal in nature, and that these 75 percent have a cost factor behind them of 90 percent of all outlay for the work comp dollar (see figure 5.1, page 69).

If you also consider the cost containment studies listed above, not only are chiropractors better trained for these types of injuries, but the cost of letting them treat the injuries is less.

All injuries should not immediately go to either a chiropractor or a medical doctor. It is up to you to assess the type of injury, then determine to whom to send the worker. When the injury warrants it, chiropractic may be the most effective, efficient, and least expensive choice, in terms of both direct and indirect costs.

Five Most Frequent Injuries
Specific Injuries

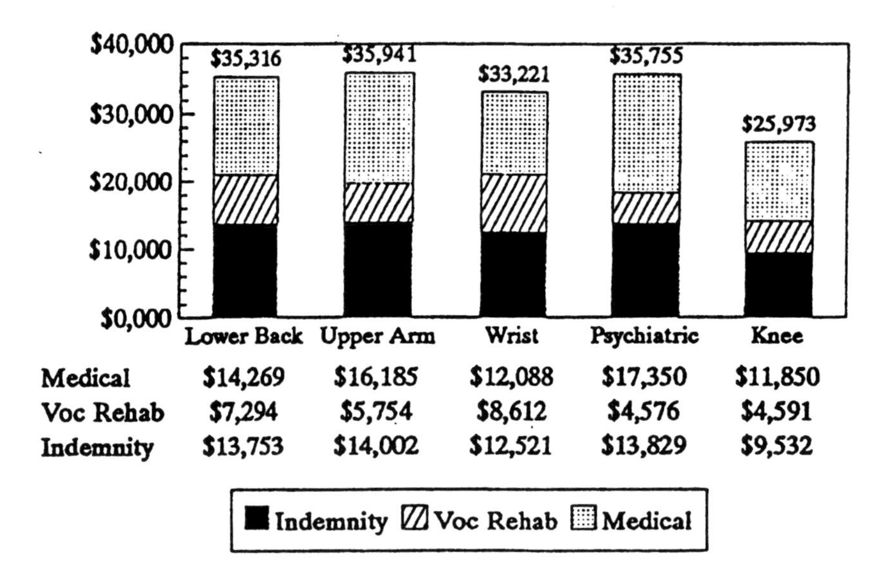

Figure 5.1: Incurred Benefit Costs: Specific Injuries
Source: M.L. Miller, *Cumulative Injuries and Specific Injuries in California, A Report to Industry.* (San Francisco: California Workers' Compensation Institute, 1995).

As a suggestion, in your Rolodex or inside the first aid kit, or even on the chart for work injuries, record the names of different medical providers with whom you have established a relationship. List them by specialty. Some examples follow:

Dr. John Smith, MD
123 Main St.
555-1654
For: cuts, broken bones, surgery, medicines

Dr. Bill Jones, DC
567 North Ave.
555-9876
For: sprains, strains, slips and falls

Why do this? In an emergency, your employees may not think clearly. This little chart or card may save time in a crisis and eliminate debate on the subject. Any worker should be able to see this information at any time; it may provide guidance on which type of care to select should the need arise. The worker may select another doctor of the same ilk and still be on the right track.

"Another study of low back disorders, the results of which were published in the June 2, 1990 issue of the *British Medical Journal*, showed that chiropractic treatment was more effective than hospital outpatient management.

"The study, which was conducted by T.W. Meade, Sandra Dyer, Wendy Browne, Joy Townsend, and A.O. Frank, concluded, 'For patients with low back pain in whom manipulation is not contraindicated, chiropractic almost certainly confers worthwhile, long-term benefits in comparison with hospital outpatient management.'

"The study was conducted in chiropractic and hospital outpatient clinics in 11 centers in Britain. It included 741 patients, ages 18 to 65, who had no contraindications to manipulation and who had not been treated within the previous month. Measures were taken by recording changes in the score on the Oswestry pain disability questionnaire and in the results of the tests of straight leg raising and lumbar flexion. A benefit of about seven percentage points was noted on the Oswestry scale when [the patients were] seen at two years. The benefit of chiropractic treatment became more evident throughout the follow-up period. Secondary outcome measures showed that chiropractic was more beneficial.

"In the British study, the high incidence of back pain was acknowledged, along with its chronic and recurrent nature in many patients, and its contribution as a main cause of absence from work. The study was done in order to evaluate largely anecdotal reports from patients, therapists and doctors of chiropractic of great improvement in back conditions after manipulative treatments." [21]

Medical Tests

Often times when dealing with different medical professionals, you will hear many different terms regarding tests that either need to be done or have already been done to your employee. Briefly, the main tests listed here are those that would be done on an employee in most doctor's offices following an injury.

MRI, Magnetic Resonance Imaging, is a test to determine soft tissue problems or tumors. This is the main test of choice for someone under the age of 50 where you are looking for something like a blown disc, or a disc bulge. This test can cost anywhere from $600-$1500, depending on what type of facility the test is being done at and the particular state you are in.

This is a very expensive test; yet if a disc problem is suspected, there is no other test that can give you the same conclusions. This test is one that should not be run as a routine screening test, but more as one that is done if examination findings warrant it. It obviously can run up medical costs, which can be of benefit to both the plaintiff and defense when litigation is involved.

It is important, when this test is recommended by the medical provider, that you talk to the doctor and ascertain the necessity for the test. The advantage here is knowing your doctor. If you have a good relationship with the doctor, chances are the test will not be done unnecessarily.

X-rays are another test needed to confirm or to determine a diagnosis. They are really only good for finding a problem with the hard tissues of the body (hard tissues are the bones). Some forms of tumors can be seen, but generally they need to be in the bone. A problem in the lungs can be seen also, but as stated earlier, generally x-rays are taken to determine problems with bones. The problems that are seen are fractures, misalignments of vertebra and other bones, some bone tumors, and problems in areas like the lung fields.

Most doctors, when an injury has occurred, will take x-rays as an inexpensive way to rule out a major problem. For areas like the neck, up to seven x-rays may be taken. The lower back can have up to five different views. Obviously, other areas will have different numbers of views. Again, by keeping in touch with the medical provider, you are able to know what has been done and what may additionally be needed.

Other, sometimes expensive, tests you should be aware of are Electromyography (EMG) and Somatosensory Electrical Potential (SSEP). These are both neurological in nature. EMG tests the muscles to see if the function of the nerve supply to the muscle is good. SSEP (Somatosensory) is a test to determine the nerve flow to a particular nerve.

During almost any examination done by the doctor, there will be any number of orthopedic and neurologic tests that can easily be done in the office. These should not have additional costs, but instead should be considered as part of the initial examination fees.

When talking to the doctor, you may hear names like straight leg raise, Soto Hall, Braggard's, Faber's, and others for orthopedic tests; and names like pinwheel, sharp and dull, dynamometer, and others for neurological tests. These are all usual and customary tests to do as early screenings of the injured worker. The details of these different tests are more medical in nature and not of great importance in this text.

Other Considerations

In regard to medical care, it should be noted that most states, if not all, have enacted laws against fraudulent workers' compensation claims. This fact is important for a few reasons. First, there is the fact that some medical clinics or individual doctors' offices have arrangements with attorneys. These arrangements allow the doctor to gain new patients, though the methods for getting these patients are not always ethical.

The second reason this is important is that by staying aware of the injured worker's status, knowing who your employees are, knowing the doctors involved, and following other preventative measures indicated throughout this book, you can avoid the many types of problems and costs that can occur after the employee has been injured.

Do your employees, the injured workers, have the right to choose the doctor they want to attend to the injury? That depends on which state you are in and whether it is part of your company's policies. (See appendix C.)

In the case of the injured worker, it is important to know the worker's needs and wants, and to either go with them or discuss them with the employee and come to some form of a compromise. By doing this, you set the whole process up as positive, instead of creating an antagonistic situation. Remember, as stated earlier in the book, the worker is concerned about a variety of issues; don't add to them by forcing the issue of which doctor to see.

Summary

Medical care is one of the most costly aspects of workers' compensation. There are many ways to help reduce your costs in this area. The first is by knowing who your company doctor or doctors are—not just by name, but also by how they are working with your employees, what their billing practices are, what kinds of tests they are running, etc. You do not need to be a doctor to get some basic information regarding the medical care of your employees. Choose the company doctor well and you can easily eliminate, or at least reduce, a large percentage of your costs.

Chapter 6

Prevention

Now that we have discussed the operations and operators in workers' compensation systems, let's discuss how to reduce injuries at your company.

This is the realm of prevention. It is highly talked about, highly debated, and of critical importance in business. Prevention does and will save your company major dollars in workers' compensation costs, employee down time, employee substitution, and training.

Prevention starts on the day the employee comes to your office for an interview. At that time, you should discuss the type of work the prospective employee will be doing, and whether he or she feels able to do the work.

Secondly, inform the applicant that there will be a doctor's exam to evaluate his or her fitness for the work in question and to detect any signs of predisposition to injury. The examination will include a medical history and basic orthopedic and neurologic tests, as well as x-rays to detect hidden problems. In some states, drug testing is permitted as a prescreening to employment.

Ergonomic factors can play a role in who is hired for which job. Since it is very difficult for a person who is 5'1" to do assembly at a 4' tall bench and for a 6' tall person to work hunched over a 3'6" bench, you might direct these applicants to other positions within the company where each could flourish, not flounder. This height disparity between the worker and the work can lead to injury.

Looking at it another way, it would be a great disadvantage to hire someone 6'6" to work all day at a table that is 3' tall. Put together a body profile for the job in question. Do measurements of the job surface or area to get a rough perspective of appropriate height for any job.

Thirdly, if the job involved requires any lifting, bending, or stooping, make sure the employee is wearing a lumbosacral support of some kind. This will in itself help to build a level of confidence in the employee about the company, as well as to protect both the worker and the company from a great majority of lower back lifting injuries.

One thing that most employers do not do is to establish some form of stretching or exercise program for the employees first thing in the morning. About 80 percent of all injuries happen within the first two hours and during the last hour of the workday.

Think about your morning routine. You get up, shower, get dressed, and get in your car and go to work. No exercise, no stretching, nothing to prepare you for the day's work. If everybody took five to ten minutes each morning and did some basic job-related stretching, then a large percentage of injuries and illnesses, both on and off the job, would be prevented. Does a football player or gymnast stretch out before a sporting event to prevent injuries? Yes. That sport is his or her job. Stretching out can help you and your staff avoid getting injured in the first place. (See figure 6.1 below.)

SHOULDER/UPPER BACK STRETCH

- **BEND knees slightly, FLATTEN back by doing "Pelvic Tilt".**
- **Clasp hands/fingers behind back, then straighten arms while looking upwards.**
- **HOLD 3-5 seconds, REPEAT 3-10 times.**

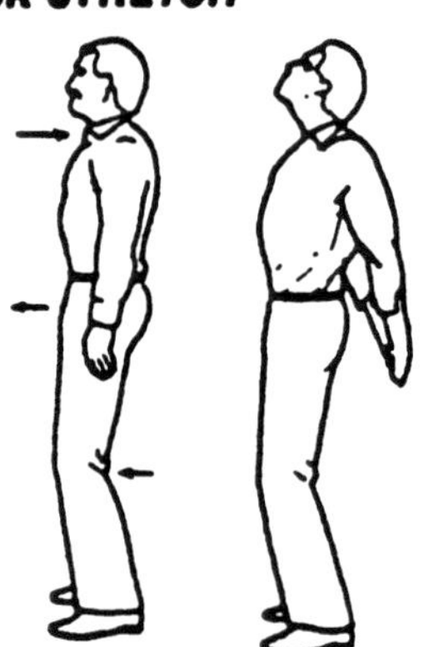

STEP FLEX

- **Place one foot on STABLE raised surface (box, palat, bumper).**
- **Rear heel FLAT.**
- **With hands on knee, bring chest TOWARDS knee.**
- **HOLD 3 seconds, REPEAT 3-10 times, then SWITCH legs.**

Figure 6.1: Exercises

Source: BACKSAFE ®

"BACKSAFE" is an injury-prevention program that helps employers reduce workers' compensation claims and expense by teaching workers how to do their specific jobs in a "biomechanically correct" way. (To contact BACKSAFE, call 1-800-775-2225.)

One thing is important to remember here: the larger the company, the bigger the workers' compensation insurance bills and the greater the overall costs. But the smaller company with a few employees can also feel the effects of having an employee out with an injury, and it has just as much, if not more, to lose in costs and job time lost as the owner of the mega corporation.

The Hiring Process

The hiring process is the beginning of the prevention process. It is here that you should explain to the prospective employee about your safety programs and accident prevention programs. Management should also give prospective employees a copy of the job description and/or office policies for them to look at.

These written documents should not only give a complete job description, but they should also contain the company's policies on drug testing, safety, medical review in the contingent job offer, along with information about vacations, sick time, raises, etc. These written policies or guidelines are important for a number of reasons. They help to initially establish the way your company does its business, and they also help your company comply with state and federal regulations regarding hiring, firing, and office policies.

Americans With Disabilities Act

The Americans with Disabilities Act (ADA) became effective in July 1992, for companies with more than 25 employees. Basically, this act restricts the employer from asking certain information of the prospective employee. It also prevents an employer from discriminating against disabled persons applying for jobs.

This act made it illegal for employers to ask questions about previous job-related injuries or claims. Employers had used these questions to screen out applicants whom they thought were likely to file more claims. Now, the employer may ask questions only about the applicant's ability to perform a certain job, not whether he or she has a disability. The act also prevents the employer from asking the applicant to perform certain types of tests that are used to screen out people with disabilities.

Bottom line for the ADA is, if the person can do the job, disabled or not, you must have a substantial reason for not hiring him; if the applicant is hired, it is up to the employer to make "reasonable accommodations" to help the disabled employee accomplish the assigned job. For example, ramps may need to be installed for wheelchair access to the workplace, equipment may need to be modified, etc. These necessary modifications do not have to be an "undue hardship" on the operation of the business.

The written job description needs to establish the basic functions of the job and must also be in line with the ADA. These guidelines are necessary to allow job applicants the ability to see for themselves what the job entails, and to then make a determination for themselves whether they think they can do the job. It also helps a doctor who may be doing a pre-employment physical. If the doctor knows the details of a specific job, he can make a much better assessment of someone's fitness for work.

Employment Applications

"Applications are solicited. All companies need to review and streamline their application forms to comply with the ADA. A good test to decide if a question is lawful is to ask, 'How is this question job-related?'

"What can be asked of the prospective employee? A simple question can be asked of every applicant, without exception: 'Do you know of any reason why you would not be able to perform the various functions of the job you are seeking?'

"The Americans with Disabilities Act requires that a conditional job offer be made prior to testing or screening applicants. Under the ADA, there *cannot* be any pre-job physicals, medical inquiries, or questions about prior workers' compensation claims. There *cannot* be job-related questions regarding agility and strength [or] about ability to perform job-related functions. For example, employees being hired to drive company vehicles can be hired contingent upon proof they have a valid driver's license. Drug testing is also permitted, if required by company policy." [22]

At this point, the employee is given a conditional job offer, contingent upon passing a medical examination. The testing must be administered to everyone working in a particular job; otherwise, you are in violation of the ADA. Questions from this point on can include previous injuries both on and off the job.

Obviously, if the prospective employee has lied or misled the employer, the job offer can then be rescinded. In addition, the ADA says that a conditional job offer may be withdrawn for two reasons: when a medical professional determines that the individual

1) is a threat to himself or others; or
2) is unable to perform essential functions of the job (with reasonable accommodation, if requested).

Injury Prevention Programs

Prevention programs establish goals for controlling losses and may use incentives to help employees attain those goals. The goals need to be realistic and there should be a concrete method for measuring progress toward these goals. In another realm, if you assigned your sales representative a goal of $1,000,000 in sales in a territory producing only $10,000 and issued only one sales report per year, the representative would have very few ways to gauge progress toward the goal. Since the sales goal is so far beyond reality, the sales representative would probably have given up trying to reach it anyway. Try to set more realistic and measurable goals. For example, if accidents are reduced 40 percent, have a company picnic. If accidents are reduced by 60 percent, spring for some steaks! Employees who know exactly what they have to do to receive a reward are much more likely to try.

It is important to keep track of the different types of losses in each department. Things to look for are:

- "frequency of lost-time injuries per number of full-time employees (frequency);
- total lost workdays per number of full-time employees (severity);
- average lost workdays per lost-time injury; and
- total costs.

"Lost workdays rapidly drive up workers' compensation losses. Research indicates that the severity of an injury—measured by total lost workdays—not frequency, is a major cause of rising costs." [23]

Company Safety Policy

The next logical step in the progression toward a prevention-based business is a safety program or policy.

The following is a small portion of the safety policy from an office manual:

> It is the policy of Smith Automotive that accident prevention shall be considered of primary importance in all phases of operation and administration. It is the intention of the office manager to provide safe and healthy working conditions and to establish and insist upon safe practices at all times by all employees. The prevention of accidents is an objective affecting all staff and their activities. It is, therefore, a basic requirement that the office manager makes the safety of employees an integral part of his or her regular management function. It is equally the duty of each

employee to accept and follow established safety regulations and procedures. Employees are expected to assist the manager in accident prevention activities. Unsafe conditions must be reported. Fellow employees that need help should be assisted. Everyone is responsible for the housekeeping duties that pertain to their jobs.

Any injury that occurs on the job, even a slight cut or strain, must be reported to the office manager as soon as possible. In no circumstance, except an emergency, should an employee leave work without reporting an injury that has occurred.

The Injury and Illness Prevention Program outlines this office's policies and procedures to maintain a safe and healthy work environment for all employees. Responsibility for implementing the Injury Prevention Program is given to Mary Smith. She will coordinate all efforts and oversee the enforcement of all office safety rules and policies.

The aforementioned example is a good beginning. All areas of safety need to be addressed in a full-length program. Part of the policy should include not only safety-based thoughts and actions, but also materials or products that can be used to improve the safety of any company, such as lower back belts, keyboard pads, wrist supports, and masks to help prevent inhalation of toxic chemicals, etc. Use only the products which apply to your business. Your secretary may not need a chemical mask unless your computer really has fumes! All of these items would obviously depend on your company's needs and its type of business.

On-The-Job Safety

Training your employees to think "safety" and to act in a safe manner is easy if you are willing to take the time. As a manager or supervisor, it is your duty to look at what is needed to prevent injuries and keep the workplace safe.

Start by looking at the job and work site.

- Is the job site or job safe?
- What are the built-in hazards of the job and the job site?
- Can any of these hazards be changed or removed?
- Is the employee the right height and build for the job?
- Do we have the right equipment for the job? Is the equipment safe?

Your next step is to look at the employees and the area around the employees.

- Are there any major distractions?
- Is there unnecessary equipment in the area?
- Can the employee get to the work area safely?

As part of the training process, it is necessary to show the employee exactly how the job gets done. Experience has shown that after an injury has occurred, employees are sometimes just thrown into another job. They are not properly trained or even shown how to do the job.

Watching how the employee is doing the job is your next step. Just because you have thoroughly explained every facet of a job and how it should be done properly does not mean that the employee will understand it. After the explanation, watching the employee will allow you to understand a couple of things. First, it lets you see if he or she understood and is able to correctly perform, and second, it lets you see how well you explained it. The level of understanding at this point is not only dependent on the worker, but it also depends on whether you have the ability to properly communicate how and why something needs to be done.

So now the employee has been trained, thinks safety all the time, knows how to do the job properly, and is shielded from any type of interruption or distraction. Is your job done? Not quite.

Now comes the easiest part of this entire process: monitoring the employees. Periodically check with them and make sure that they are still doing things the way they should be done.

- Have they become lazy about their jobs?
- Are they just going through the motions?

Remember that training is an ongoing process. Just because you have trained employees once, don't expect them to be perfect all the time. You need to do safety checks; make sure that there is appropriate information on the walls, such as "lifting" posters, "what to do if" posters, and "how to" posters; and make sure the workers are using assigned equipment such as lifting belts, masks, and keyboard pads. (See figure 6.2 on page 80 for an example of a safety poster.)

After you observe the people (your employees), you also have to inspect the equipment to make sure it is being kept up properly.

One company in San Diego has gone one step further to reduce its costs and keep its job site safe. The owner of the company was getting tired of paying over $85,000 per month for his workers' compensation insurance premiums, and having numerous large claims against it. The claims for 1992 were more than $1.5 million. He instituted as part of his safety program a bonus system based on accident reduction in the workplace. He basically started paying his employees for not having injuries. In less than one year, he cut his claims from $1.5 million to less than $200,000, and the number of injury claims from over 1,000 to less than 150.

TO KEEP THOSE KINKS OUT OF YOUR BACK!

RIGHT

1. SQUAT DOWN
2. BEND KNEES
3. BACK STRAIGHT
4. ARMS CLOSE
5. KEEP LOAD CLOSE

CLAIMS FOR CHIROPRACTIC SERVICE ARE APPROVED UNDER WORKMAN'S COMPENSATION LAW FOR ON THE JOB INJURIES. CHIROPRACTIC HAS EARNED AN ENVIABLE RECORD OF RESULTS IN INDUSTRIAL INJURIES, ESPECIALLY BACK AND NECK INJURIES AND ALLIED STRAINS. STATISTICS SHOW THAT LESS WORK TIME IS LOST UNDER CHIROPRACTIC CARE THAN OTHER HEALTH METHODS. THIS IS AN IMPORTANT BENEFIT TO BOTH EMPLOYER AND EMPLOYEE.

DOCTORS NAME

ADDRESS AND PHONE

DISPLAY IN A PROMINENT PLACE

Figure 6.2: Safety Poster on "Lifting"

The owner did this legally. I have heard of some companies doing this by not filing claims and doing other things to avoid making the claims, which is unethical. This company began an intensive safety program, and by following the ideas mentioned previously was able to make dramatic changes.

As stated earlier in this text, find a medical professional whom you can trust and who is willing to help your company with ergonomics, safety, and prevention. Nearly all chiropractors are willing to go out to businesses and do ergonomic studies, show employees how to lift, and teach other safety steps necessary for the prevention of injuries. Most of them do this work for no charge.

Good Work Practices

The following lists of work place do's and don't's, and the list of guidelines for moving and lifting, have been reproduced from Ilene Stone's *Ergonomics: A Basic Guide*.

Work Place Do's

- "Position lights and work to avoid glare and shadows.
- Use matte finishes on surfaces, and remove shiny objects.
- Wear proper clothing and required personal protective equipment.
- Know the symptoms of heat or cold illness and how to deal with it.
- Take rest breaks in areas with moderate temperatures.
- Allow time for the body to adjust to new temperatures.
- Eat and drink appropriately.
- Maintain good physical fitness.
- Ensure that noise levels are acceptable.
- Maintain equipment properly to minimize noise and vibration.
- Use good work/rest schedules.
- Adjust chairs properly.
- Use chairs that suit the task.
- Use footrests, where necessary, to support the feet and take the pressure off the back of the legs.
- Keep shoulders relaxed.
- Hold elbows comfortably at the sides of the body.
- Keep wrists straight.
- Move joints within a reasonable range.
- Change body position frequently to ease strain on muscles and joints.
- Adjust work surface and seat properly.

- Place frequently used items in front of the body and within reach.
- Use platforms where necessary.
- Use footrests or rails.
- Wear appropriate footwear.

Work Place Don'ts

- Don't cramp knees in small spaces.
- Don't stand for a long period of time without a break.
- Don't twist the back, neck, or wrists.
- Don't lean forward or sideways too far.
- Don't crouch for too long.
- Don't work with arms raised for too long." [24]

Guidelines for Moving and Lifting

- "Move loads of appropriate size, weight, and shape only.
- Reduce the weight of the load.
  - Reduce the size by repackaging.
  - Reduce the number of objects carried at once.
  - Assign more people to lift heavy loads.
- Make the load easier to handle.
  - Ensure good footing.
  - Change the load's size and shape to move the center of gravity nearer to the lifter.
  - Use spring-loaded bins to raise the load to at least hip height.
  - Use more than one person or a mechanical device to move the load.
  - Drag or roll the load with handling aids such as carts, ropes, or slings.
  - Transfer the weight of the load to stronger parts of the body using a hand grip, straps, or belts.
- Use storage techniques to ease material handling.
  - Store the load at or above hip height to avoid the need to bend.
  - Use wall brackets, shelving, or pallet stands of appropriate height.
  - Load pallets so that heavy articles are around the edges of the pallet and not in the center.
  - Remove shelves from their brackets with forklifts, load at a proper height, and then return shelves to their brackets.

- Minimize the distance a load is carried.
  - Improve the layout of the work area.
  - Relocate production or storage areas.
- Minimize the number of lifts.
  - Assign more people to the task.
  - Use mechanical aids.
  - Rearrange the storage or work area.
- Reduce twisting of the body.
  - Keep all loads in front of the body.
  - Allow enough space for the entire body to turn.
  - Turn by moving the feet rather than twisting the body." [25]

As a manager or employer, it is vitally important that you keep prevention in mind while you are hiring, training, and dealing with your employees or co-workers.

Proper Equipment

Proper equipment is of utmost importance for you to be able to keep the costs of workers' compensation in check. By using preventative equipment such as lower back belts, wrist supports, and typing or keyboard accessories, you will be able to not only reduce work comp claims, but also to keep morale and employee confidence at its highest point. This, in turn, will benefit your company because you will have higher productivity.

When you hire that new employee, make sure you have the best possible equipment for that employee to use. The better the equipment you start the employee off with, the less the chance of an injury occurring.

You cannot eliminate all injuries just by having good equipment, but you can definitely decrease the number of injuries you are currently dealing with. If you owned a construction business, you wouldn't give your workers inferior drills or hammers; you would find the best tools so that the work could be done properly and efficiently, with the minimal amount of exposure to possible injury.

Why not provide the best secretary's chair, a proper wrist appliance for data entry personnel, or a good lower back belt for employees who perform repetitive lifting? It really is the same idea. You give the employee the best tools to do any type of job.

Providing your secretary, office worker, or data entry person with the best ergonomic chair helps prevent soreness in the lower back and gives better support to the back, neck, knees, and wrists, and thus improves morale and productivity. When you think about your employees, especially the office staff, realize that they are basically in one position for six to eight hours every day. This repetitive position is the beginning of a cumulative trauma claim.

This area of ergonomics needs to be addressed by all business people. It is not only the beginning of all work comp claims, but it is also a place where businesses can save great deals of money in indirect costs. If your business is short-sighted and only addresses the initial expenses of trying to implement an ergonomic cost-containment program, then your business is going to lose the battle. If, on the other hand, your business can look at the long-range benefits from this area of implementation, you can clearly see that when you add up the costs of fewer lost days of work, smaller outlays for temporary employees, and/or less disruption of work schedules, the benefits of good equipment easily outweigh the initial costs.

Summary

When looking at how your business can cut costs, you will find that planning ahead is probably one of the easiest ways to reduce overhead and other costs. Train your employees right from the beginning and keep monitoring their progress. Give them the safety equipment that you would use if you were performing the job at hand. Take five minutes in the morning before shifts begin and institute a short stretching or exercise program. These are the things that should be done by all businesses, regardless of the size or scope of the business.

Ergonomics

When ergonomic applications are being considered, it is the office workers who are most often overlooked. These employees sit all day, usually in poorly designed chairs at work stations that are often incorrectly lit, with their keyboards at the wrong height. The drawing in figure 7.1 (on page 86) shows what the proper work station for a data processor or secretary should look like, and what the proper position of the employee should be.

Later in this chapter you will find additional data about seating, work stations, and employee evaluations of work stations. If you apply this data, you can prevent injuries at your workplace—injuries that don't have to happen.

To give you a good idea of how important it is to take care of the seated employee, see figure 7.2 (page 87), which shows how much weight is loaded onto the lower back intervertebral discs, depending on trunk position and whether the person is supported or unsupported.

Golden Rules for Office Chairs

The following is a small set of rules for office chairs:

1. Office chairs must be adaptable to both traditional office work and to video display work stations.

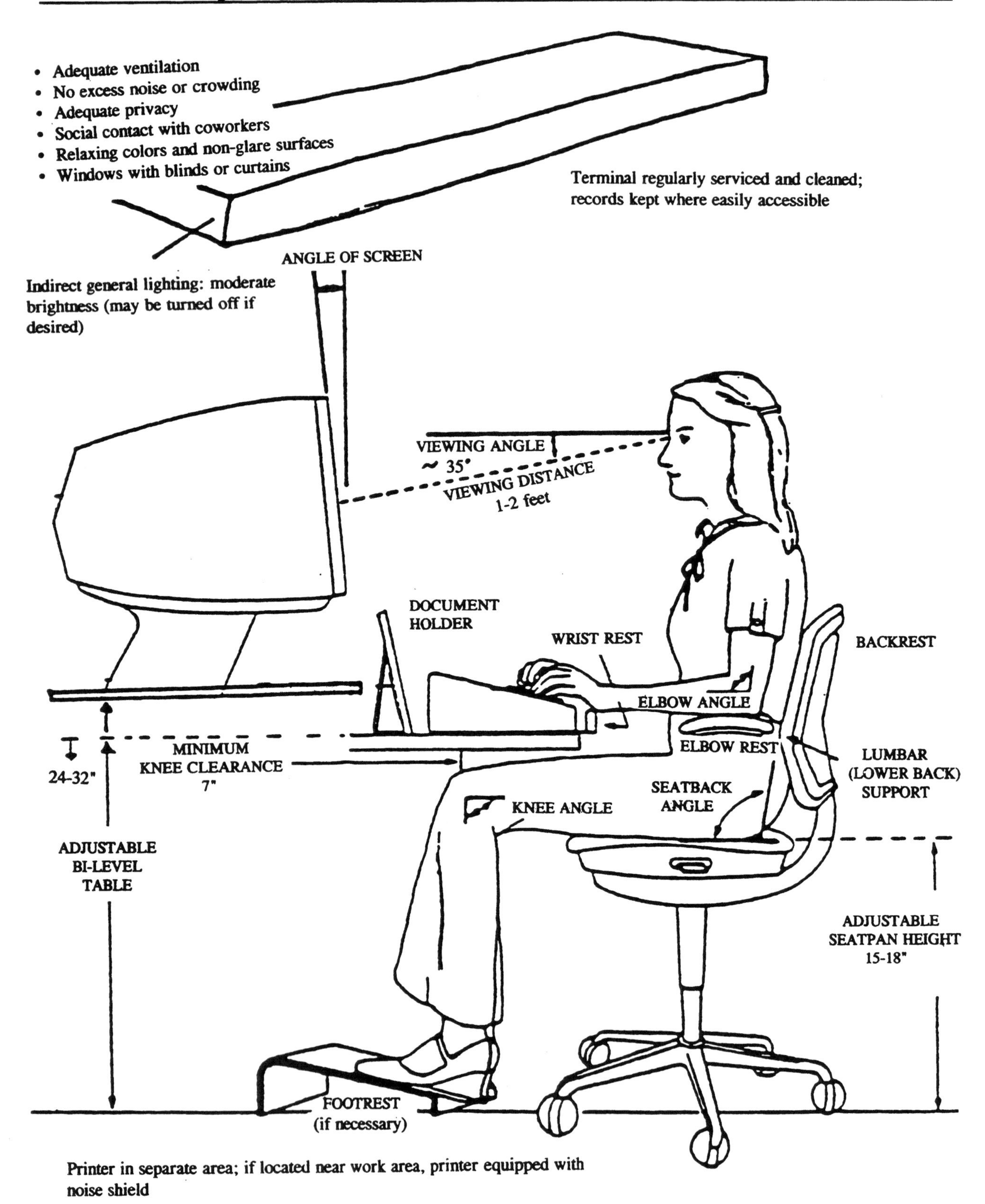

Figure 7.1: Proper Seating and Office Set-Up

Unsupp.	100 kg	120 kg	125 kg	131 kg	189 kg	210 kg
Supp.	91 kg	88 kg	10 kg	97 kg	137 kg	117 kg

Discal loads of an average subject (weight 70 kg and height 170 cm) with different trunk postures with unsupported and supported upper limbs.

Figure 7.2: Weight of Person Sitting in Different Positions
Source: Occhipinti, et al.

2. Office chairs must be able to be used in both an upright and a reclined position.
3. The backrest must be able to be moved back and forth.
4. A backrest height of 24-27 inches vertically above the seat surface is an ergonomic necessity.
5. The backrest must have a well formed lumbar pad.
6. The seat pan should measure 20-23 inches across and 18-22 inches from back to front.
7. Footrests are important.
8. Chairs must fulfill all modern requirements:
   - adjustable height (19-25 inches)
   - rounded front edge of seatpan
   - five-arm base with castors or glides
   - user-friendly controls.

The area of office ergonomics and VDTs (video display terminals) is a major concern to all businesses, as syndromes like carpal tunnel and VDT stress and other forms of repetitive motion injuries are on the increase. Ergonomics is now the fastest growing area of workers' compensation injuries. It is also the easiest area in which to help prevent injuries simply by following some of the aforementioned strategies and spending a few dollars. Remember, it is easier and less expensive to prevent injuries than to pay for them after they happen.

Work-site, seating, and job-design evaluations are very important ways to prevent both one-time type injuries and cumulative trauma injuries. The following sections contain different types of evaluations that can be used in almost any type of business.

Using Checklists in Work-Site Evaluations

In *Workbook for Ergonomic Considerations in Office Design*, Robert Arndt discusses the use of checklists for work-site evaluations. The following information describes how to use checklists when evaluating office chairs, and includes samples of a checklist and a questionnaire suggesting what sort of questions you should ask your employees.

"One of the most straightforward and simple methods of conducting a work-site evaluation involves the use of checklists. The basic purpose of a checklist is to provide structure and guidance in carrying out an evaluation. It can never be substituted for a complete and thorough evaluation. One of its primary limitations is its inability to consider other potential interacting factors. Thus, none of the specifications should be considered as rigid requirements; there can always be exceptions.

"The format of each checklist should be set up in such a way that it identifies the important factors and characteristics that should be considered. The checklists should [seek] an answer to a question. Each of the questions should be phrased in such a way that a response of 'yes' is desired (unless the question is not applicable).

"It must be kept in mind that the purpose of the checklist is to draw attention to the item so that it is given consideration. With repeated use of the lists, the user will eventually realize that some items are more important than others. In addition, it may be desirable to delete or add items to the existing lists.

"The data collected for each checklist will also provide a[n] historical record of conditions that existed at the time of the original assessment." [26]

Instructions for Using the Checklist for Chairs

On page 90, you will find a sample checklist showing the type of information you should include when constructing your company's own checklist for evaluating chairs. You should keep in mind the considerations listed below to determine what kinds of questions your checklist should include. Once you have decided what questions to ask, you can then use your company checklist "to evaluate present chairs, present needs, or proposed chairs. The appropriate chair[s] will depend on the population of users, the task, and the work environment. The evaluation of chairs should ordinarily be supplemented by surveys of workers. [See page 91 for a sample of a worker questionnaire.]

- "If frequent adjustments are necessary due to differing tasks or multiple users, check 3.

- If the same chair is to be used for a variety of different tasks, check 13, 14, 15, 16, and 24.
- If the same type of chair is to be used by individuals varying greatly in stature, check 4, 8, 11, 12, 15, and 19.
- If the user must sit for long hours in the chair, check 5, 9, 10, 14, 18, 19, and 20.
- If the chair is used with fixed height working surfaces, check 4 and 16.
- If the chair has armrests, check 14, 15, and 16.
- If the backrest of the chair tilts, check 8 and 10.

Height

The range of heights necessary will depend on the size of the population of users and the working height. All workers should be able to place their feet flat on the floor or on a footrest.

- 90 percent of women: 14-17.5 inches
- 90 percent of men: 15.5-19.2 inches

Backrest

High backrests are preferred by most workers. The need for up-and-down and forward/backward adjustments will depend on the size of the chair and the range of worker sizes. If the backrest tilts, it must also lock or be adjustable in tension.

Armrests

If they are intended to support the arms while working, they may have to be adjustable in height and distance apart. Armrests for most clerical jobs should be optional since they may interfere with work.

Seatpan

If the seatpan tilts, it should be either lockable or adjustable in tension in order to provide good support. The tilt angle is largely a matter of worker preference.

Comfort

Comfort is subjective and can be evaluated only by asking workers. Comfort may be affected by the individual characteristics of workers, the nature of the task, and the environment." [27]

Checklist for Chairs [27]

PRESENT YES	PRESENT NO	PROPOSED YES	PROPOSED NO		
					A. Height
[]	[]	[]	[]	1. []	Is the range of height adjustment adequate?
[]	[]	[]	[]	2. []	Can the chair height be easily adjusted?
[]	[]	[]	[]	3. []	Can adjustment be made from the seated position?
[]	[]	[]	[]	4. []	Are adequate footrests available?
					B. Backrest
[]	[]	[]	[]	5. []	Does the chair have a high backrest?
[]	[]	[]	[]	6. []	Does the backrest interfere with arm movements?
[]	[]	[]	[]	7. []	Is the lumbar support adequate?
[]	[]	[]	[]	8. []	Is the tension of the backrest adjustable?
[]	[]	[]	[]	9. []	Does the backrest tilt back?
[]	[]	[]	[]	10. []	Does the backrest lock in position?
[]	[]	[]	[]	11. []	Can the backrest be adjusted up and down?
[]	[]	[]	[]	12. []	Can the backrest be adjusted forward?
					C. Armrests
[]	[]	[]	[]	13. []	Does the chair have armrests?
[]	[]	[]	[]	14. []	Are armrests appropriate for the job?
[]	[]	[]	[]	15. []	Are armrests optional?
[]	[]	[]	[]	16. []	Do armrests interfere with movement?
					D. Seatpan
[]	[]	[]	[]	17. []	Does the seatpan have a rounded front edge?
[]	[]	[]	[]	18. []	Does the seatpan tilt?
[]	[]	[]	[]	19. []	Is seatpan tension adjustable?
[]	[]	[]	[]	20. []	Does seatpan lock in position?
					E. Safety
[]	[]	[]	[]	21. []	Is the chair stable?
[]	[]	[]	[]	22. []	Does the chair have a five-leg base?
[]	[]	[]	[]	23. []	Are casters matched to the floor?
[]	[]	[]	[]	24. []	Can casters be changed?
[]	[]	[]	[]	25. []	Are all adjustments safe against self or unintentional release?
[]	[]	[]	[]	26. []	Does the chair meet all applicable fire codes?
					F. Comfort
[]	[]	[]	[]	27. []	Is the chair adequately padded?
[]	[]	[]	[]	28. []	Are materials appropriate?
[]	[]	[]	[]	29. []	Is the chair comfortable?
					G. Other
[]	[]	[]	[]	30. []	Can the chair be easily maintained?
[]	[]	[]	[]	31. []	Can maintenance be performed in the field?

Questionnaire for Worker Evaluation of Chairs [27]

1. Your name ______________________________

2. Name of chair ______________________________

3. How long do you use this chair? __________ hours [per day].

4. Did you make any adjustments before or while using the chair? ___ yes___ no.
   If yes, please describe ______________________________
   ____
   ____

5. Please rate the following features of the chair by circling a number for each.

	very comfortable			very uncomfortable	
a. chair height	1	2	3	4	5
b. clearance for feet and calves under chair	1	2	3	4	5
c. seat comfort	1	2	3	4	5
d. backrest comfort	1	2	3	4	5
e. overall chair comfort	1	2	3	4	5
	very good			**very bad**	
f. maneuverability	1	2	3	4	5
g. ease of adjustment	1	2	3	4	5
h. size of chair	1	2	3	4	5
i. general appearance	1	2	3	4	5

6.What do you like best about this chair? ______________________________
____
____
____

7.What do you like least about the chair? ______________________________
____
____
____

8. Describe the type of work you are performing while using this chair.

typing	_____	hours [daily]
VDT use	_____	hours [daily]
clerical	_____	hours [daily]
other	_____	hours [daily]

Environmental Hazards

The following information concerns the types of environmental hazards that can be found in the workplace, and contains suggestions on how to deal with them.

Lighting

"Poor lighting can contribute to:

- accidents and injuries
- tired, sore eyes
- headaches
- red and dry eyes
- blurred vision
- double vision

Reflections from surrounding walls, furniture, and equipment; the contrast between objects and the background; and natural light levels affect lighting. Common lighting problems include too much or too little light, glare, and shadows. Glare and shadows can force the worker to use awkward body positions to see work. Lighting problems can cause accidents.

Adjustable task lighting provides more light where needed. Avoid glare and flicker by properly positioning and maintaining fluorescent lights.

Temperature

"Humidity and air movement affect the perception of temperature. Working in cold temperatures can cause:

- discomfort
- chills and shivering
- loss of finger and limb dexterity
- frostbite
- hypothermia

Working in hot temperatures can cause:

- headaches
- dizziness
- sweating, rashes, and dehydration
- weariness

- muscle spasms in legs or abdomen
- fainting
- unconsciousness

Temperature problems can be reduced by:

- decreasing air velocity and blocking drafts in cold areas.
- increasing air flow and blocking sources of heat by using heat shields in hot areas.
- limiting time and level of exposure through job design, rest breaks, and work schedules.
- providing rest areas with moderate temperatures.
- providing protective clothing and equipment.

Noise

"Working with noise can cause:

- workers not to hear conversations and warning signals
- disturbance of concentration
- social isolation
- body position problems from adopting an awkward position so the ear can face the direction of the expected sound cues
- hearing loss

Noise problems can be reduced by:

- replacing noisy equipment with quieter machinery.
- using sound-dampening devices, such as mats, under motors.
- isolating noisy equipment.
- limiting exposure time.
- wearing hearing protection.

Vibration

"Segmental vibration involves vibration of parts of the body, such as the hand or arm. It can be caused by hand-held power tools. Whole body vibration involves vibration of the entire body.

Working with vibration can cause:

- tingling and numbness
- pain in fingers
- motion sickness
- fluttering of the soft tissue of the neck and face
- blurred vision
- tingling sensations
- difficulty with balance
- fatigue
- pain in specific areas, such as the lower back

Vibration problems can be reduced by:

- using dampening materials to absorb vibration.
- using good seats on vehicles.
- using good work/rest schedules.
- maintaining equipment properly." [28]

The following are two sample surveys to demonstrate what types of questions to ask your employees to determine what hazards they might be facing in the workplace.

Questions for Field Evaluations of Worker Problems [29]

Health Symptoms Survey

The following symptoms concern the systems and functions of your body. Please try to answer each question by indicating how often you have experienced each of the described problems within the past six months: (Circle one)

	Never	Sometimes	Frequently	Constantly
Pain or stiffness in your arms	1	2	3	4
Pain or stiffness in your neck	1	2	3	4
Pain or stiffness in your shoulders	1	2	3	4
Pain or stiffness in your back	1	2	3	4
Burning eyes	1	2	3	4
Eyestrain	1	2	3	4
Headaches	1	2	3	4
Leg cramps	1	2	3	4

Work Environment Problems Survey [30]

Indicate whether any of the following conditions create problems in your job by circling the appropriate answer.

	Never	Sometimes	Frequently	Constantly
Temperature	1	2	3	4
Drafts	1	2	3	4
Odors	1	2	3	4
Equipment noise	1	2	3	4
Distracting noises	1	2	3	4
Chair comfort	1	2	3	4
Backrest comfort	1	2	3	4
Workspace	1	2	3	4
Storage space	1	2	3	4
Legroom	1	2	3	4
Table height	1	2	3	4
Keyboard height	1	2	3	4
Place to rest arms	1	2	3	4
Amount of light	1	2	3	4
Glare from lights	1	2	3	4
Glare from windows	1	2	3	4
Reflections on desks	1	2	3	4
Reflections on VDTs	1	2	3	4

Specifications for Work Surfaces

The following information should be considered when designing the work stations that your employees will be using.

Height

- "Fixed work surfaces should have a height of approximately 29 inches for most tasks and 27 inches for keying tasks.
- All adjustment ranges are dependent on the worker population, the task, and the equipment. The following are therefore only very rough estimates.
  - Adjustable tables should have a range of 26-29 inches.
  - Adjustable keyboard supports should have a range of 24-28 inches.
  - This can vary considerably depending on the design of the keyboard and the adjustment range of the chair.

Reach

- About 95% of all adults can reach 22-26 inches without bending over.
- Reaches exceeding 30 inches will necessitate that many workers stand up.
- Work can be performed most efficiently within about 16 inches of the body.

Leg Room

Minimum leg room allowed is about 27 inches deep and 27 inches wide. The height should be sufficient to allow the legs to be upright while the worker is sitting in the chair (feet on floor or footrest).

Computer Work Space

Tables designed for computers should allow the display surface to be positioned at least 18 inches from the front of the keyboard platform." [31]

Guidelines for Working Positions

It is important for employers to establish good working positions for their employees. You should also obtain responses directly from your employees so that you can stay in touch with their needs and alleviate their concerns (the surveys on pages 94-95 can give you ideas of what to ask them). Here are some guidelines for good working positions:

- "In a well-designed workplace, the worker has the opportunity to choose from among a variety of well-balanced working positions and to change among them frequently.

- Working tables and benches should be adjustable. Adjustability for working height is particularly important in order to match the work station to the worker's individual body size and particular task. Adjustability ensures that the worker can carry out work in well-balanced body positions.

- If adjustability is not built in, there need to be other ways of matching the height of the work station with the worker's dimensions. Platforms to raise the shorter worker or pedestals on top of work stations for the tall worker should be considered.

- Space organization is another important aspect of workplace design. There should be enough room to move around and to change body position. Providing built-in foot rails or portable footrests allows the worker to shift body weight from one leg to the other. Elbow supports for precision work help reduce tension in the upper arms and neck. Controls and tools should be located so that the worker can reach them easily without twisting or bending.

- Where it is possible, a seat should be provided so that the worker can do the job either sitting or standing. The seat must place the worker at a height that suits the type of work being done. For work that requires standing only, a seat should be provided in any case to allow the worker to sit occasionally. Seats at the workplace expand the variety of possible body positions, giving the worker more flexibility.

- The benefits from greater flexibility and a variety of body positions are twofold. The number of muscles involved in the work is increased, equalizing the distribution of loads on different parts of the body. [As a result] there is less strain on the individual muscles and joints used to maintain the upright position. At the same time, changing body positions improves the blood supply to the working muscles. Both effects contribute to the reduction of overall fatigue.

- Proper flooring materials are also a major factor in standing comfort." [32]

Job Design

"Job design and workplace design go hand in hand to ensure the health and safety of workers. Job design is the [manner in which] a set of tasks, or a job, is organized. Job design determines:

- what tasks are done
- how tasks are done
- how many tasks are done
- the order in which tasks are done

Job design is a continual process. The idea is to make adjustments as conditions change within the workplace. Good job design accommodates workers' mental and physical characteristics by paying particular attention to:

- muscular energy expenditure
- physical force requirements and body position
- stress

Good job design minimizes muscular fatigue and keeps energy expenditures within the limits of workers' abilities. The basic principle is to balance static and dynamic work. Static work, such as holding a box at arm's length in front of the body, tenses or contracts muscles. Maintaining a fixed and sometimes awkward posture is physically demanding, requires energy, and reduces the flow of blood to the muscles. Dynamic work, such as digging a trench or scrubbing a floor, involves rhythmic contraction and relaxation of muscles. It also encourages blood flow, which removes the body's waste products and provides the necessary oxygen and nutrients. But overuse of any body part leads to pain, strain, and possible injury.

Good job design keeps physical force requirements reasonable. Body positions that maximize strength while maintaining safety minimize the demands put on the musculoskeletal system. Excessive or hazardous demands increase bending, twisting, overreaching, awkward body positions, lifting and carrying, and tasks with repetitive motions.

Good job design prevents harmful stress. Stress is the body's nonspecific response to any demand made upon it. It allows the body to tap energy stores needed to meet physical or mental demands. [But] not all stress is harmful: some stress is necessary to motivate and challenge individuals. Positive stress occurs when the demands of the job match the worker's skills and knowledge with the equipment and materials needed to do the job. Positive stress gives the worker a sense of achievement and self-esteem.

Characteristics of job design that determine whether stress is harmful or beneficial are:

- amount of control the worker has over tasks
- level of responsibility
- level of participation in task-related and general workplace activities
- amount and type of supervision
- extent to which roles are understood
- extent and nature of social support systems." [33]

The following page contains a sample checklist with ideas of the types of questions you should ask your employees. You should construct a checklist specifically for your company and distribute it to your employees. This way, the employees can provide you with specific feedback on any problems they might be having, which will give you an idea of where changes or improvements need to be made.

Checklist for Job Design Hazards [34]

Posture

Body - bent, twisted, or unsupported?	yes ___ no ___
Neck - bent, tilted, or twisted?	yes ___ no ___
Shoulders - hunched?	yes ___ no ___
Arms - elevated?	yes ___ no ___
Wrists - bent or twisted?	yes ___ no ___
Legs - pressure on back of knees or thighs?	yes ___ no ___

Movement

Joints - at extremes of their range of movement?	yes ___ no___
Grips - difficult to grasp, too much strength required?	yes ___ no___
Direction - movement awkward and not natural?	yes ___ no___
Pushing or pulling - tasks that require great effort?	yes ___ no___
Lifting - heavy or awkward loads?	yes ___ no___

Duration of Movements

Repetitive tasks - same body parts used over and over?	yes ___ no___
Static positions - few or no opportunities to change position?	yes ___ no___
Fast work pace - muscle tension and stress?	yes ___ no___

Work/Rest Schedules

Long work period - fatigue?	yes ___ no___

Features of Good Job Design

"Good job design:

- minimizes energy expenditure and force requirements.
- balances static and dynamic work.
- allows some decision making so workers can vary activities according to personal needs, work habits, and the circumstances in the workplace.
- gives workers a sense of accomplishment.

- includes training so workers know what tasks to do and how to do them.
- provides good work/rest schedules.
- allows an adjustment period for physically demanding jobs.
- provides feedback to the workers about how they are doing.

Steps Toward Good Job Design

"Achieving good job design involves administrative practices that determine what the worker does [and] for how long, where, and when. Here are the steps that an occupational health and safety committee [in your company] should advocate to managers, supervisors, and workers:

1. "**Vary body positions and movements**. Avoid static body positions and repetitive movements to alleviate boredom. Design jobs with varied tasks that require changes in body position, muscles used, and mental activities. Two methods of doing this are job enlargement and job rotation.

Job enlargement redesigns jobs to include more and different tasks. This increases the variety of motions and the time between repeated motions. Job enlargement adds interest to the work and may give workers more responsibility.

Job rotation moves workers from one job to another. It distributes the load of stressful jobs among a group of workers. It also reduces the time each worker spends in one position and the number of repetitions of the same motion. Job rotation can provide relief only if the various tasks use different muscle groups. An example is a job that allows the worker to sit instead of stand.

2. "**Optimize work/rest schedules**. All work schedules, especially rotating shift work and the extended workday schedules, must be well designed. Rest breaks help alleviate the problems of unavoidable repetitive movements and static body positions. Frequent short breaks are sometimes preferable to fewer long breaks. Encourage workers to change body position and exercise during rest breaks. It is important that workers stretch and use different muscle groups.

3. "**Allow for an adjustment period**. When work demands physical effort, have an adjustment period for new workers and for all workers after holidays, layoffs, or illnesses. Allow [for] time to become accustomed to the physical demand of work by gradually 'getting in shape.'

4. "**Providing training**. Training in correct work procedures and equipment operation is needed so that workers understand what is expected of them and how to work safely. Training should be organized, consistent, and ongoing. It may occur in a classroom or on the job.

5. "**Ensure that there are enough workers to complete tasks safely**. Avoid injuries by having more workers help with tasks, such as lifting.

- Determine the number of workers needed according to the load's weight and size.
- Identify one worker as the leader who will plan and explain the entire move.
- Have the leader position workers carrying the load to ensure good weight distribution.
- Ask workers to raise and lower the load on the leader's command.

6. "**Vary work tasks to vary mental activities**. Coordinate tasks so that they are balanced. Balance the workload among workers." [35]

Summary

Many employers think that by reducing workplace risks to their warehouse workers, laborers, and construction crews, they have eliminated the major causes of workers' compensation accidents. Many employers don't think that "simple secretarial work" can cause injuries. Unfortunately, they are wrong on both counts. The fastest growing segment of cases involves repetitive motion injuries (e.g., carpal tunnel syndrome and VDT stress), the vast majority of which are preventable by applying ergonomics. If the equipment (chair, work table), the job site, and the job design are correct for that worker, you will reduce your risk of having an expensive cumulative trauma workers' compensation case filed against you. Take the time now to use these checklists, listen to what your workers say about the chairs, desks, lighting, noise, and temperature conditions under which they work. Simple readjustments to the office thermostat, reducing glare and noise, and buying the office furniture which allows adjustments to meet each user's needs, will save dollars in future workers' compensation claims, lost work days, and productivity.

Chapter 8

Disability and Rehabilitation

The term "disability" generally means a loss or reduction of physical or mental function that causes an individual to have a reduced ability to earn a living.

There are several different types of disability: permanent total, permanent partial, temporary total, and temporary partial.

Permanent total disability (PTD) and permanent partial disability (PPD) are exactly what they sound like. The worker is permanently unable to engage in substantially remunerative employment. Simply, this means that the worker is unable to work in any job realm that will give any reasonable amount of salary. Examples would be someone who lost both arms or was blinded by a work injury.

PTD compensation will continue until one of the following things happens:

- the claimant returns to substantially remunerative employment;
- the claimant dies; or
- the claimant exhausts (where applicable) the maximum dollar limit set by statute.

Temporary total or temporary partial disability (TTD or TPD) is also just like it sounds. When a worker is injured and must miss work for a couple of weeks for recovery, he or she is entitled to some form of compensation while off of work. It is temporary in nature, and payable during the acute post-injury phase while the employee is going through medical care or recovery.

In workers who have sustained lower back injuries, what are the factors of long-term disability? They are numerous and widespread.

Many factors have been shown to increase the nature and severity of a lower back injury, and the degree of the nature and severity of the injury directly affects the length of treatment time and disability. The primary factors shown in research for the frequency of severe injuries and that of reinjury are ergonomic occupational factors.

There have been numerous studies done that show that occupations involving heavy lifting have the highest incidence of lower back injuries. The excess risk for lower back pain among workers engaged in heavy lifting is focused among those who lift heavy objects, especially when the objects are held away from the body while lifting and when bending and twisting occurs while lifting.

Another factor in this issue is repetitive lifting. When an employee is in an occupation that requires repetitive lifting, there is also a higher risk of lower back injury, even at levels of lifting considered to be light to moderate. It has even been shown that jobs which require frequent lifting of objects weighing 25 pounds have seen increased incidence and risk. And on the other side of the coin, workers who did little to no lifting, such as secretaries or government workers, were least likely to be afflicted with lower back injuries.

Another factor of long-term disability for lower back injuries is prolonged sitting and vibration. As a result of numerous studies, prolonged sitting is believed to increase the risk of lower back pain.

Through the research that has been done, the following factors have been established as having the highest ergonomic risk factors for lower back injury:

1. "Excessive loads
2. Manual handling tasks
3. Certain body movements
4. Prolonged sitting
5. Vibration." [36]

When we get down to the basics on disability, it becomes evident that you as the employer need to have some idea as to how long your employee could be out of the work environment, and what the chances are of your employee ever returning to work.

Many researchers have noted that the prognosis for patients with acute back pain is excellent. The percentages cited show that from 75 to 80 percent of lower back pain cases will be resolved without complication within 30 days and will account for only 7 to 13 percent of the total workers' compensation costs for lower back injury.

The Quebec Study states that 74.2 percent of compensated workers with spinal disorders in Quebec were absent from work for less than one month. However, 7.4 percent of compensated workers lose more than six months from work-related lower back injury claims.

"The Weyerhaeuser study done in 1986 found that workers off [of] work for a period of more than six months have only a 50% chance of returning to gainful employment, and if off [of] work for one year the percentage of returning is 25%. After two years, the percentage plummets to 0.0%." [37]

"The chart below (figure 8.1) gives a good representation of the breakdown of different types of disabling work injuries." [38]

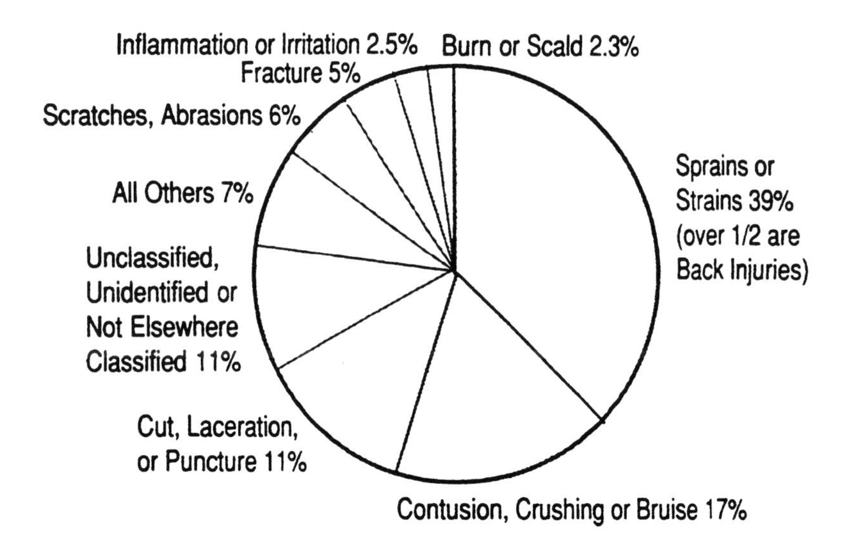

Figure 8.1: Disabling Work Injuries by Nature of Injury
Source: Canadian Centre for Occupational Health and Safety, *Nature of Injury by Occupation.* (Toronto: Canadian Centre for Occupational Health and Safety, 1985).

Return to Work, Rehabilitation, and Work Restrictions

The most critical aspect of disability, rehabilitation, and return to work is the way the injured worker is treated. The injury should be treated as legitimate, and the worker should be reminded of his need to return to work. It should also be stressed that the company is behind the worker.

As the employer or manager, it is important—both to your company and to the worker—that the employee return to work as quickly as possible. This will help reduce costs. One sure way to help the workers' compensation problem that presents itself today is for the worker to return to work quickly; thus, once again the employee becomes a productive member of the company and society, instead of being "a helpless victim" of the system.

All companies need to establish a well-thought-out return-to-work and rehabilitation program. Typically, most insurance companies are required to set up rehabilitation programs, but it is a mistake to rely on the insurance company exclusively. Work out a plan yourself. Contact rehabilitation companies and actively participate.

The longer the employee is off of work, the tougher it is to get him or her back to work. If the employee is sitting at home, collecting a paycheck and not working, it can become very easy to not want to return to work. The mental attitude of these employees can also cause a problem. They are home, feeling bad, and are told by a doctor that they cannot work for a while. Staying out of the work environment and thinking about how bad they feel tends to propagate the sense that they are not able to do anything. This tends to make them feel worse for a longer period of time and makes the process of getting them back into the workplace that much tougher.

As we all know, when you have time on your hands, not only can you get into trouble, but you also begin to feel every ache, pain, and symptom, regardless of how insignificant they may be. After a while, the employee begins to feel that he or she has no choice but to remain on disability. All the employee has to look forward to at this point is benefit payments or litigation.

One thing that employers are concerned with is, if they bring an employee back to work, must that employee return to the same job he or she had before the injury? In most states, this is not the case. When that employee returns to work, he must return to a job that is within his abilities and the limitations set forth by the medical provider. Although you may still need someone else to do that worker's job if he is only allowed to do light duty work for a while, you as the employer will still benefit from this. You will be saving dollars, because although the employee has not recovered enough to do his own job, he can do light duty work and someone else can take over his job while he recovers. This will not only facilitate the work getting done, but you will not need to hire someone outside of the company to do the employee's regular work in addition to paying benefits to the injured employee. This results in both time and money being saved.

The off-of-work employee is collecting benefits. However, the difference between what you pay him and what he collects from disability amounts to a wage loss for the employee. If the employee can return to some form of work—light duty or some other job—you can reduce the amount of wage loss the employee may currently be experiencing, thus giving him incentive to come back to work.

Allowing the employee to return to work, any kind of work, is vital to both the employee and the company. It is very costly to have an employee out of work for any length of time. If that person is able to do some job, you will cut your costs by his returning to work. This is the benefit to having some type of light duty option in the workplace.

Not only does the employee return to work, but it helps his self-esteem, shows him that returning to work is a possibility, and also shows him that the company is concerned about him. It really doesn't matter if the work is part-time, clerical, or whatever; it is important to have some type of program set up for these injured workers.

It is important to note, also, that the Americans with Disabilities Act prohibits an employer from discriminating against a person with a disability, as defined by the Act, who is "qualified" for a desired job. Therefore, you cannot refuse to let an individual with a disability return to work because the worker is not fully recovered from injury, unless the worker (1) cannot perform the essential functions of the desired job without accommodation; or (2) would pose a significant risk of substantial harm that could not be reduced to an acceptable level with reasonable accommodation. Since reasonable accommodation may include reassignment to a vacant position, an employer may be required to consider an employee's qualifications to perform other vacant jobs as well as the job held when injured.

When the injured worker returns to the job site, it is important to not just throw him back into the job. Sit down with the employee the day before his return to work and go over the doctor's recommendations regarding the worker's capabilities. Explain to the worker that the company is behind the return and will do anything it can to make the transition back to work as easy as possible. Also, make sure that all supervisors or managers have been briefed on the employee's return and limitations in the workplace.

In setting up the return to work, it is important to look for ways to enhance the workplace without demoralizing the employee. Make sure that the job considered for the return is a job that shows a contribution to the company, not just a "busy work" type of job.

Another important step in the return-to-work process is to set up some time limits for the light duty. Rehabilitation is an integral part of this process. As the employer, you may or may not know how to approach this. In most states, the insurance companies handle most of this process. They set up the rehabilitation, put the injured worker through it, and then either retrain the worker for different work or institute a work hardening program to allow the worker to return to his previous job. (For a breakdown of the various states' requirements, refer to appendix D.)

What is work hardening? It is a program that puts the workers through individual training to get their bodies prepared to return to the job force. It is not sit-ups or leg raises; instead, it is workers pushing wheelbarrows up a ramp, or other types of activities that mimic their jobs. Usually the employees will participate in a work hardening program for about four to six weeks; then they should be able to return to their previous work.

Disability and rehabilitation are areas that a lot of employers have many suspicions about. They are usually expensive, decrease productivity, and generally cause problems at the worksite. As you have read, it doesn't have to be that way. Working with the employee, medical provider, and the insurance company can make the whole process quite easy to deal with. Bring the employee back to work as soon as possible, set up work hardening programs if your company is large enough, and/or go to rehabilitation centers to see what is being done. This will give you and your company a better perspective on the rehabilitation process.

The main comment to make here is that if you do everything you can to prevent the injuries, deal with the injured worker as a human being, and make efforts to get the worker back to work quickly, the whole process will seem a lot simpler.

Understanding Work Restrictions

When your employee returns to the workplace, with the doctor's release, it is important to remember that you are required to follow the restrictions placed on the worker's return. These restrictions are usually laid out in the return-to-work slip from the doctor (see figure 8.2 on page 109), but the terms may be quite confusing.

What is "light work"? What does "no heavy lifting" mean to a construction worker? Does it apply differently to a model than to a professional wrestler? You have to make arrangements for the employee to be able to return to his job in accordance with the Americans with Disabilities Act and the restrictions placed upon the employee by his medical provider.

Terms like "sedentary work", "light work", "heavy work", or even "very heavy work" can be both confusing and misleading. The following are definitions of these terms:

Sedentary work: Lifting 10 pounds maximum and occasionally lifting and/or carrying such articles as dockets, ledgers, and small tools. Although a sedentary job is defined as one which involves sitting, a certain amount of walking and standing is often necessary in carrying out job duties. Jobs are sedentary if walking and standing are only occasionally required and if other sedentary criteria are met.

Light work: Lifting 20 pounds maximum with frequent lifting and/or carrying of objects weighing up to 10 pounds. Even though the weight lifted may only be a negligible amount, a job in this category requires walking or standing to a significant degree, or involves sitting most of the time, with a degree of pushing and pulling of arm and/or leg controls.

ATTENDING DOCTOR'S
RETURN TO WORK RECOMMENDATIONS

Company Name

PATIENT'S NAME (First) (Middle Initial) (Last)

DATE OF INJURY /ILLNESS

DIAGNOSIS

TO BE COMPLETED BY ATTENDING DOCTOR - PLEASE CHECK

I saw and treated this patient on ________ and:
Date

1. ___ Recommend patient return to work with no limitiation on ________
Date

2. ___ Patient may return to work capable of performing the degree of work checked below with the following limitations:

DEGREE	LIMITATIONS
___ SEDENTARY WORK. Lifting 10 pounds maximum and occasionally lifting and/or carrying such articles as dockets, ledgers, and small tools. Although a sedentary job is defined as one which involves sitting, a certain amount of walking and standing is often necessary in carrying out job duties. Jobs are sedentary if walking and standing are required only occasionally and other sedentary criteria are met.	1. In an 8 hour work day the patient may: a. Stand/Walk ___ None ___ 4-6 Hours ___ 1-4 Hours ___ 6-8 Hours b. Sit ___ 1-3 Hours ___ 3-5 Hours ___ 5-8 Hours c. Drive ___ 1-3 Hours ___ 3-5 Hours ___ 5-8 Hours
___ LIGHT WORK. Lifting 20 pounds maximum with frequent lifting and/or carrying of objects weighing up to 10 pounds. Even though the weight lifted may be only a negligible amount, a job is in this category when it requires walking or standing to a significant degree or when it involves sitting most of the time with a degree of pushing and pulling of arm and/or leg controls.	2. Patient may use hands for repetitive: ___ Simple Grasping ___ Pushing & Pulling ___ Fine Manipulation
___ MEDIUM WORK. Lifting 50 pounds maximum with frequent lifting and/or carrying of objects weighing up to 25 pounds.	3. Patient may use feet for repetitive movement as in operating foot controls: ___ Yes ___ No
___ HEAVY WORK. Lifting 100 pounds maximum with frequent lifting and/or carrying of objects weighing up to 50 pounds.	4. Patient is able to: Frequently / Occasionally / Not at all
___ VERY HEAVY WORK. Lifting objects in excess of 100 pounds with frequent lifting and/or carrying of objects weighing 50 pounds or more.	a. Bend ___ ___ ___ b. Squat ___ ___ ___ c. Climb ___ ___ ___
Other instructions and/or limitations	

3. ___ These restrictions are in effect until __________ or until patient is re-evaluated on ______.
Date Date

4. ___ Patient is totally incapacitated at this time. Patient will be re-evaluated on __________.
Date

Doctor's Signature | Date

AUTHORIZATION TO RELEASE INFORMATION

I hereby authorize my attending doctor to release any information or copies thereof acquired in the course my examination or treatment for the injury identified above to my employer or his representative.

Patient's Signature | Date

Figure 8.2: Sample of a Doctor's Return to Work Form

Source: Parker Foundation for Chiropractic Research

Medium work: Lifting 50 pounds maximum with frequent lifting and/or carrying of objects weighing up to 25 pounds.

Heavy work: Lifting 100 pounds maximum with frequent lifting and/or carrying of objects weighing up to 50 pounds.

Very heavy work: Lifting objects in excess of 100 pounds with frequent lifting and/or carrying of objects weighing 50 pounds or more.

This may help clarify the matter. But what about the 5'1" secretary who hurts her back at work and gets sent back to work with a "no heavy work" restriction? Is there anything else you have to do? Does her job need to be modified?

In this case it does not, unless there are times when she is occasionally lifting 100 pounds. In the case of a construction worker who hurts his leg and is restricted to light work for three weeks, however, there may be modifications needed to be able to return him to his job. The beams he is carrying may weigh more than 20 pounds, and thus he cannot do the same type of work he was doing prior to the injury. You will have to find him some sort of work he can do that will accommodate the restrictions instructed by the doctor.

Rehabilitation comes in where work restrictions leave off. If the employee is unable to return to former job duties, he or she is entitled to be rehabilitated or retrained. This training takes into consideration the employee's work restrictions, abilities, and even intelligence. The employee does not go through an IQ test—the rehabilitation company looks at what the worker is basically able to do if properly trained.

An example would be if a construction worker ruptured a disc in his lower back. He obviously could not return to doing construction work. He would go to rehabilitation with a restriction something like "no lifting over 10 pounds with no repetitive bending, lifting, or stooping". After this, some basic testing would be done to determine what other skills he may have. Following the testing, a rehabilitation counselor will begin the process of training.

You won't find that a laborer with no professional skills will be trained to be a lawyer or a doctor. The retraining will be done within all limits set up by the doctor. For instance, the construction worker might be retrained to do computer programming, secretarial work, or some other low level physical work. This is how the system of rehabilitation works.

Summary

So far, so good. The injured worker has been treated by the right doctor. You have analyzed and improved safety and ergonomics in your workplace to eliminate the causes of the accident, trusting that through your efforts it won't happen again. Now, how long will the employee be off of work and what are the chances of this person returning to work?

Talk to the treating doctor and the injured worker. In the interim, there may be part-time, light duty work which the employee can do in the company. The employee's feeling of self-worth improves and your wage-loss decreases. It's a win-win situation.

The doctor's opinion should be included, whether or not there is hope for the employee's full recovery (and return to the original job). If there is hope, work hardening to physically prepare the worker for his previous job is a viable option. If return to the old job is not a possibility, you or the insurance company may engage the services of a rehabilitation firm or counselor to ascertain what other gifts, talents, or abilities this worker has, and then commence training the worker for a brand new career. You may find that the injured assembly worker can type 80 words per minute with no errors and outwork your entire secretarial staff; or that the secretary with carpal tunnel may be the new "diamond in the rough" in your sales department. Your employee returns empowered by new skills, confident that your company really wants him back, and knowing that you cared enough to make an effort to help when the need was overwhelming.

As with medical care, costs can be contained if you communicate with the rehabilitation counselor. You may not want someone trained as a customer service representative if you don't have that job in your company. If the counselor has a clear job description of all the positions in your company that fit the doctor's restrictions, then the worker's training will prepare the worker directly without unnecessary delaying steps. For example, if you have a customer service department using IBM PC's, you'd like the employee trained on an IBM PC compatible, not a Macintosh. Training the employee on a Macintosh, while well and good, doesn't help you get the employee back into the office sooner. Costs are relatively easy to control when you make the effort.

Chapter 9

Workers' Compensation Strategies

When you examine a diamond, you will find that the larger facets are fairly easy to see. But to appreciate the beautiful power of the whole gem, you have to consider the side facets too. In this chapter, we'll consider the side facets of the workers' compensation system and present problem-solving techniques for dealing with fraud, work days lost, increases in workers' compensation premiums, the insurance company itself, and just plain horror stories. Certain problems are basically inherent in the system.

Fraud

Fraud is a major problem of the workers' compensation system, in any state, any community, and any place. It is your job as the employer, manager, or owner to do whatever you can to help prevent the problems.

What is fraud? Is it medical mismanagement? Is it an employee claiming a false injury? Is it an employer claiming no injury happened? All of these can be cases of fraud; they can also be cases of misinterpretation, miscommunication, or misunderstanding.

Did that employee really get hurt? You as the employer need to be sure, either way. If it did happen and you just did not see it, do not automatically assume it might be a fraudulent claim. Check it out! Do your own investigation.

The insurance company will do some investigating, but the reality is that it doesn't matter to them if it is fraud or not. If the claim costs them a lot of money, they will simply raise your rates or make your company put up some form of deposit, just in case. We all know, they will get their money.

Doing an in-house investigation can save you a lot of money, time, and aggravation. But in order to discover whether the injury is real or not, you have many things to consider.

First, know your employee. If the worker has been with the company for many years with no problems or complications and is a good worker, the chances of him filing a phony claim are limited. Then again, if the employee started last week and is already making a claim, you need to be suspicious. It still doesn't mean fraud, only caution.

Secondly, look at the supposed incident. Has your company ever had someone injured in that way before? Is it possible that the injury could have happened the way it was reported by the employee? This could be something that ergonomic studies, in advance, could have prevented.

Thirdly, ask questions of the other workers in the same area. Did they see anything? Were there any other signs of problems?

Once the injured employee has seen a doctor, and may be off from work, you have other things to look at. Does the employee seem concerned about her job? Is the employee available for phone calls to handle problems? These are signs that the injury probably did happen.

When you call the employee and never get anyone home, or if the employee doesn't seem to care about the job, especially in the early stages of the claim, be on the lookout for fraud. Ask other employees who might be friends of the injured worker whether or not she plans to come back to work. Find out what she has been doing.

An example of a work comp claim gone bad and costing the employer a great deal of money happened to one of my patients who owns a small business. One of his workers claimed an injury. The employee went for medical care and everything seemed okay. After the worker was off from work for about two months, the employer became very suspicious, so he began an investigation of his own. He hired private investigators to photograph and take movies of the worker. His intention was to find fraud, report it to the insurance company, have the employee held liable, and save both himself and the insurance company money.

Well, here is how it turned out. The employee, who had a lower back injury and could not bend or lift anything over 10 pounds, was photographed doing roofing work (not his usual job), waterskiing, playing football, and many other things that would be nearly impossible for him to do had there truly been a lower back injury.

This information was presented to the insurance company, but after their investigation, they basically said to the employer that they did not care. They awarded the employee rehabilitation at a cost of over $25,000, plus disability benefits, lost wages, and a settlement of nearly $20,000. The employer screamed, yelled, and hollered about this to everyone he could. He kept trying to stop the claim and was stonewalled at every turn.

After all this, the insurance company then raised the company's workers' compensation insurance premiums by 15 percent, to nearly pay back the $25,000 the insurer had paid out.

Really horrific. And this is not the only example available. This was a case that may have started out okay, but then escalated into the workers' compensation claim from hell. The main thing to keep in mind, however, is that even though a few cases could turn out to be major headaches, the majority of workers' compensation cases can be worked out in a reasonable manner if you stay informed and keep in close contact with all parties involved.

Fraud in Medical Care

Medical care is another way the system can be defrauded. Investigative news programs like "Dateline" and "60 Minutes" expose medical cost overruns in the workers' compensation system fairly often. Prior to publication of this book, a crew from "Prime Time" opened a small medical office in Los Angeles. Then, with the aid of someone previously convicted of defrauding the system, they waited for the onslaught of individuals making numerous promises of how many patients they could bring in and how much money the clinic could make.

On the videotape of this scenario, you could see many doctors and non-medical people come into the office offering to refer patients in exchange for money. These people get their "patients" off the streets and out of unemployment lines and encourage them to file workers' compensation claims. "Prime Time's" reporter went on to say that if the clinic had been real and all the "runners" (those people who illegally acquire patients for lawyers and doctors) had produced the patients they said they could, the clinic would have grossed over $20 million a year, all from fraudulent cases. This is one of the ways that medical care providers can get involved in work comp fraud.

Most of this kind of fraud happens because of the way the laws are written. In California, prior to July 1993, for one workers' compensation case there could be as many as seven or eight medical legal reports, each costing $750-$1500. (See figure 9.1 on page 116.) With this kind of incentive, it is very easy to understand why this kind of fraud takes place. It is hoped that the new laws in California will put a plug in this rather large hole.

In general, with medical fraud, it is not the doctor in the field as much as the med-legal clinics who do most of the evaluations. Remember, not every clinic is fraudulent; it is a very small percentage that really is.

Med-Legal Evaluation Billings, Payments, and Disputes Per Report
Specific and Cumulative Injuries

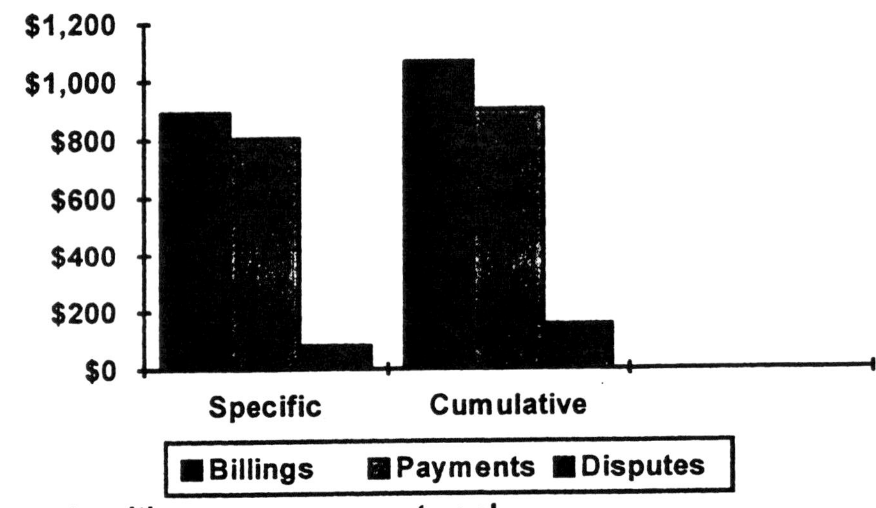

Reports with nonzero payments only

Figure 9.1: Med-Legal Costs Chart
Source: M.L. Miller, *Cumulative Injuries and Specific Injuries in California, A Report to Industry.* (San Francisco: California Workers' Compensation Institute, 1995).

Another example of problems that can occur is the following. Here is a synopsis of the remarks by a Workers' Compensation Appeals Board Judge regarding a case involving Johnny Unitas, a former football star. He had filed a workers' compensation claim against his last professional team, the San Diego Chargers, for continuous trauma injuries. If you are a doctor or a legal professional, this is the kind of report that you look at because it lets you know about possible precedents or some of the more oddball cases. It allows for a little better understanding of the workers' compensation law and how it is applied.

CASE NAME: JOHN UNITAS v. SAN DIEGO CHARGERS

ZENITH INSURANCE COMPANY

CASE NO.: SDO 136983 (2 cases)

ISSUE: Employer must have knowledge of work-related disability before any duty of notification of benefits arises.

FINDINGS OF FACT: The causes of action filed herein are barred by Statute of Limitations. Defendant is not estopped to plead the Statute of Limitations. All other issues are rendered moot.

OPINION ON DECISION: Applicant, John C. Unitas, born May 7, 1933, Application for Adjudication of Claim on June 15, 1989, alleging injury to his lower back in approximately August 1973 while employed as a football player by the San Diego Chargers, then insured by Zenith Insurance Company (SDO 136983). A second application was filed on June 15, 1989, alleging injury to applicant's right shoulder in approximately August 1973 while employed as a professional football player by the San Diego Chargers, then insured by Zenith Insurance Company (SDO 136983). A third application was filed on July 23, 1987, alleging injury to "multiple parts including but not limited to internal and orthopaedic" during the period of January 1, 1973, to December 31, 1973, while employed as a professional football player by San Diego Chargers, insured by Zenith Insurance Company.

A hearing was held on all three cases on February 25, 1991. One of the issues raised in all three cases was the Statute of Limitations and estopped to raise the Statute of Limitations defense. A review of the facts presented and the applicable law lead to a finding that all three matters herein are barred by the Statute of Limitations.

As to applicant's right shoulder and lower back injuries (SDO 136983), the evidence was clear that these applications were not filed until June 15, 1989, which was almost two years after the applicant retained counsel and obtained a medical evaluation from the doctor clearly finding applicant's back and shoulder condition related to applicant's injuries while employed by the San Diego Chargers. Applicant was well aware of the industrial nature of these injuries but delayed filing applications for almost two years.

Applicant's continuous trauma case is also barred by the operation of the Statute of Limitations. Under Labor Code 5405, an application to commence original proceedings for disability indemnity must be filed within one year of the applicant's date of injury. For cumulative injuries, like the one claimed here, the "date of injury" is the date upon which the employee first suffered disability therefrom and either knew, or in the exercise of reasonable diligence should have known, that such disability was caused by his employment (Labor Code 5412). In the case herein, applicant's testimony indicated that he had aches and pains to the parts of the body claimed on the application and that he knew that these were due to playing football, as he had no other activity which would have caused his problems. Applicant had disability and knowledge that his disability was industrially related as early as 1973. Applicant's claim is clearly barred by Labor Code 5405.

Applicant contends that defendant is estopped from asserting the Statute of Limitations because of their failure to comply with the notice provision of the Labor Code. In the present case, applicant was employed by the Chargers during the period of January 1, 1973, through July 31, 1973. Labor Code 5402 did not contain a notice requirement, i.e., it did not impose a duty on the employer to notify the employee of his rights to workers compensation in case of injury. Further, the employer herein had no knowledge of applicant's assertion of a continuous trauma claim until the claim was filed in 1987. The Administrative Director's Rule 9816, however, provided for notice where the employer had "notice or knowledge of injury which: (a) required hospitalization or results in disability of more than seven days, or (b) results in death" (Cal. Admin. Code, tit. 8, 9816). Under the evidence presented herein, the employer was not put on notice of injury and was not required to give applicant any notice whatsoever Pursuant to Rule 9816 and the law as it existed in 1973.

Applicant argues that the principles in *Buena Ventura Gardens v. WCAB* (1975), 40 CCC 47 be followed. However, the principle remains that the employer must have knowledge of work-related disability before any duty of notification of benefits arises.

The evidence is clear that the applicant's continuous trauma claim is barred by the Statute of Limitations and that the defendants are not estopped from raising the statute as a defense.

All other issues are deemed moot. There are no sums available against which to award an attorney fee.

ORDER: It is ordered that applicant take nothing by reason of the Application for Adjudication of Claim filed herein.

PETITION FOR RECONSIDERATION: Filed by applicant.

Although what you just read may sound like a lot of legal mumbo jumbo, it is presented to show you some of the things that an employee can do, with the aid of a reasonably good workers' compensation attorney.

Second Injury Funds

This type of fund allows employers to hire previously injured workers, including those who may have permanent disabilities, without assuming workers' compensation liability for injuries incurred prior to employment.

As an employer, when you begin the interview process, you will probably come across individuals that have a previous disability or physical challenge. If you take them at face value, your first reaction may be to not hire these people because you don't want to put yourself in a position of potential liability. This is where you may be wrong.

Besides the obvious legal problems of discrimination, you are making the common mistake of assuming that if these employees get hurt you will be liable for their entire disability. You need to know that you are not responsible for the whole disability. Your responsibility begins at the point where the initial disability ends.

As an example, you interview a person who tells you that he has a disability of "no heavy work". This is equivalent to approximately 15 percent permanent partial disability. You choose to hire him anyway because of his other qualifications.

Two months into the job, the handicapped employee improperly lifts a box and gets injured. He or she goes through the whole workers' compensation process and is determined to have a 20 percent disability. As the current employer, your responsibility toward this disability is only 5 percent, or the difference between the employee's previous disability and his current one. While your company is not completely devoid of responsibility, the level to which your company must compensate the employee is greatly reduced.

This simple scenario lets you understand that there is nothing to be afraid of when a handicapped or partially disabled person comes to your business to apply for a job. It is also not automatic that if you hire handicapped individuals, they will be more prone to an on-the-job injury.

The only additional qualification that needs to be addressed here is, what is considered to be an injury or disability within each particular state? In some states, like California, a disability is only a disability if it has been documented by a medical examination and then written and posted within the workers' compensation realm, or sometimes the personal injury realm. (For additional information on how your state views this issue, please see appendix E.)

Dealing with the Insurance Company

The next area that needs to be addressed is how to deal with your insurance company. Initially, specific forms need to be filled out by your company, usually with the employee's statement and signature.

For most companies, this is the end of their dealings with the insurance company—except, of course, to pay the monthly or yearly premiums—but it should not be. It is your company's responsibility to stay in touch with the insurance company to monitor how things are progressing. As the employer, and as the insured, you have the authority to tell the insurance company to pay medical claims, disability benefits, rehabilitation costs, and other costs previously addressed.

Typically, the employer files the appropriate paperwork, then steps back and lets the insurance company do what it wants. Sometimes this is to your benefit, but most of the time, your involvement is not only important, but it can also save you great sums of money and aggravation, especially when the insurance bills come in.

The sample scenario of what my experience has shown is as follows. The injury happens, employee and employer fill out paperwork. The injured worker goes to the urgent care facility, where he is given some medical care. The insurance company then begins to dictate what happens from there. They say whether the injured worker can see additional medical providers, sometimes regardless of what the company doctor says.

The employee then gets aggravated by the poor treatment and seeks the counsel of an attorney. The attorney files specific paperwork which allows the worker to see whatever doctor he chooses. This incites the insurance company, which then requests independent medical examinations in order to either find a loophole so they don't have to pay additional benefits, or to make it look like the worker is not really injured; if this is not possible, they try to make the injury and residual subjective and objective complaints look much more minor than they really may be.

The worker's attorney then does the same thing, only he is trying to make the injury look worse so that the benefits can be greater. At this point the entire process is costly for the employer, insurance company, and attorney, and extremely time-consuming for all involved.

In the case of a simple, easily defined injury, your input and supervision are not very critical; the question is, what are the predominant types of injuries in your business? If your major injuries are simple, your involvement can be limited. But a large percentage of the injuries can get complicated very quickly and very easily. A portion of the escalation is due to the typical employer's lack of control, another part is due to insurance company intervention, and the remainder is due to an employee or an unscrupulous lawyer trying to take advantage of the system.

By maintaining control, or at least keeping an eye on claims, you can obviously have an effect not only on costs, but also on how well the medical intervention is being dealt with. As has been stated throughout this book, the more open and available you are with your injured employees, the smaller the chance they will want to exploit either you or the insurance company. Keep the lines of communication open with your employees. Let them know their importance to the company, regardless of how menial their jobs may be. Let them know they are needed back as part of the team, not as just more bodies on the assembly line. These simple actions can save you enormous time, money, and effort in dealing with claims and insurance companies.

How can the insurance company exploit the situation? It is not as hard as you may think. Experience has shown that in a large number of cases, you can have a "gung-ho" adjuster trying to do a good job for the insurance company. The adjusters are trying to save money, so they make the assumption, whether accurate or not, that the injured worker or the doctor might be trying to make a fraudulent claim. The adjuster holds up the doctor's payments, or the benefits to the injured worker. This then causes the employee to get an attorney to deal with the uncooperative insurance company.

This simple act alienates the employee and starts the cycle of bad feeling. Even though you had nothing to do with the problem, the employee looks at you because it is your insurance company that is denying his injury and costing him money. It becomes very easy to see how a negative situation can begin.

You, as the employer and insured, need to be in contact with the insurance company to make sure things are going smoothly. It is your responsibility to see that the injured worker is receiving the appropriate benefits and medical care. If the insurance company tells you that they see a problem, you have the right to tell them that there is no problem, and that they should pay the bills or benefits.

You would be amazed at how employees look at this situation. When you stand up for them, they tend to get well faster, get back to work quicker and, in the long run, either save you a lot of money or not cost you as much.

In some cases, no matter what you do, nothing is going to stop the inevitable progression of a case if the employee's only concern is to make the big score. The only way to deal with this is through your initial screening process.

In the case of a bad employee, however, the insurance company can be your best friend. If you suspect fraud, let them know. They have lawyers that are there to represent you, without direct cost to you. This is the case where you should fight.

Summary

As the employer, your injured (and resting at home) employee may seem like a liability, what with the financial problems associated with finding a substitute for that person, the looming thunder clouds of a large claim advancing over the horizon, and visions of your workers' compensation premiums edging higher once again. Don't be discouraged; there are proactive steps you can take to control the size of the claim and speed the employee's return. By keeping lines of communication open between you, the employee, the treating doctor, and the insurance company, you can make sure you're all on the same page in the playbook and that you are all working toward the same goal: a healthy employee. In these days of one rip-off after another, whether it is in electronics or government contracts, it is tempting to doubt the veracity of a worker's claim, but don't. Help the injured person through this difficult process of healing physically while keeping afloat financially. If it really is a substantiated case of fraud, alert the insurance company and they should handle it. In the vast majority of cases, it is what it appears to be—the person you hired is now injured and needs help. Don't be afraid to help; don't be afraid to care. The genuine concern you show could save you the escalated costs of a workers' compensation case gone bad.

Appendix A

Cost Containment Studies

Reprinted with permission from the American Chiropractic Association, *Chiropractic State of the Art*, pp. 32-35, 39-40. Washington, D.C., 1991-1992.

Oregon

The Medical director of the Workers' Compensation Board of the State of Oregon released the results of a similar study, "A Study of Time Loss Back Claims, 1971," which showed that of claimants treated by no physician other than a chiropractor, 82 percent of these workers resumed work after one week of time loss. These claims were closed without a disability award. However, of claimants treated by M.D.s, where the diagnosis seems comparable to the type of injury suffered by the workers treated by the D.C., only 41 percent of these workers resumed work after one week of time loss. See Martin R.A., "A Study of Time Loss Back Claims: Workmen's Compensation Board", Medical Director's Report, State of Oregon, Archives of the California Chiropractic Association Vol. 4, No. 1, 1975.

In a separate study of statistical information furnished by the Oregon Workers' Compensation Board in 1971, a 24-month study limited to back-injury cases involving only sprains and strains found $298.52 to be the average total cost under care of M.D.s, as compared to only $72.92 average total cost under care of D.C.s. These average total costs include doctor and hospital costs plus compensable time loss.

In yet another study of Oregon Workers' Compensation Insurance claims (October 1974 through May 1975), Bergemann and Cichoke, using strict testing criteria, obtained the following results:

	Case	Work Days Lost	Treatment Length (Days)	Treatment Costs
D.C. Care	113	18.88	53.24	$181.48
M.D. Care	114	41.16	97.38	$327.30

Kansas

A study of Kansas workers' compensation records in 1972 of average time loss and treatment costs for back injuries handled by doctors of chiropractic and medical doctors revealed the following:

	Time Lost	Treatment Costs (excluding hospital bills)
M.D. Care	13.1 days per case	$117.61 per case
D.C. Care	5.8 days per case	$68.43 per case

Iowa

A comparison of the cost of D.C. versus M.D. treatment is found in two Iowa studies covering the years 1966 and 1969. Average cost per case for M.D. treatment in 1966 was $118.74 and in 1969 was $210.86, as compared to average cost per case for D.C. care in 1966 of only $68.24 and only $79.28 in 1969. Statistical information was furnished by the Iowa Workmen's Compensation Service, 1969 and 1971. A more recent analysis in 1978 of non-operative back and neck injury claims processed by the Office of the Industrial Commissioner reported that both the average period of disability and the average amount of compensation awarded were lower for chiropractic patients than for medical patients. Appearing below is a comparison of average length of disability and average compensation cost per case, experienced by claimants of various age groups in that study:

	Chiropractic		Medical	
Age Group	**Disability (days)**	**Compen-sation Paid**	**Disability (days)**	**Compen-sation Paid**
16-24 yrs	21.1	$204.16	25.3	$319.20
24-44 yrs	22.4	$284.79	24.4	$353.62
Over 44 yrs	21.7	$258.61	27.1	$423.34
All Ages	21.9	$262.21	25.1	$360.06

California

In December 1972, C. Richard Wolf, M.D., completed a study designed to compare time loss to industrial back injury when treated by either a D.C. or an M.D., using the records of the Division of Labor Statistics and Research (Doctor's First Report of Work Injury). In summary, 1000 employees with industrial back injuries were questioned about the time lost and residual pain from injuries suffered; 629 of them responded to the questionnaire. Of the injuries reported, one-half had been treated by M.D.s and one-half by D.C. physicians. No degree of bias in study design could be determined, and there were no apparent, major identifiable differences in the two groups with regard to age or employee categories. The major differences determined were:

	Employees Treated by M.D.s	Employees Treated by D.C.s
Average lost time per employee	32 days	15.6 days
Employees reporting no lost time	21.0%	47.9%
Employees reporting lost time in excess of 60 days	13.2%	6.7%
Employees reporting complete recovery	34.8%	51.0%

Montana

In 1978, a comparison study on chiropractic and medical ambulatory care of back strain and sprain injuries in Montana from 1975 to 1978 was made. The average period of disability and average amount of compensation paid to chiropractic and medical patients with back strain and sprain injuries were as follows:

	D.C. Patients		M.D./D.O. Patients	
Age Group	**Disability (weeks)**	**Compen-sation Paid**	**Disability (weeks)**	**Compen-sation Paid**
16-24 yrs	3.0	$378.35	3.3	$466.71
25-44 yrs	1.8	$248.60	3.0	$409.83
Over 44 yrs	3.8	$352.93	4.0	$561.66
All ages	2.8	$315.93	4.0	$463.66

Wisconsin

In 1978, Daniel J. Dully, M.B.A., in Market Research, University of Wisconsin, conducted a comparative study of M.D. vs. D.C. treatment of closed industrial back injury cases for 1977. The results were as follows:

Costs	Cases	Days Lost	Treatment
M.D. Care	430	21.8	$267.58
D.C. Care	212	13.2	$145.64

A Comparison of Chiropractic, Medical, and Osteopathic Treatment Costs in Workers' Compensation Cases

In early 1988, the Foundation for Chiropractic Education and Research (FCER), with the cooperation of the Florida Department of Labor and Employment Security, Division of Workers' Compensation, completed a comprehensive analysis of Florida's workers' compensation claims for back-related injuries. The analysis compares the cost of treatment for patients of chiropractic, medical, and osteopathic doctors, and includes costs of both hospital and non-hospital services for back-related injuries.

FCER established rigid guidelines for both the data that were requested from the state of Florida and the statistical analysis which followed in an effort to overcome criticism that previous workers' compensation studies relied on selective and questionable sampling procedures.

Data for this analysis were compiled from all cases reported to the state of Florida during its 1985-86 fiscal year. The analysis covered the costs of all major diagnostic and treatment procedures and hospital services by type of physician. The analysis also compared surgical and non-surgical costs.

Highlights from the study's findings include:

- **Chiropractic patients had the lowest rate of incurring compensable injuries when compared to medical or osteopathic patients.**

  Of all patients with back-related injuries—52,091 including those who may have required surgery—33.0 percent had compensable injuries. However, the percentage of patients who incurred a compensable injury varied significantly by service provider: 36.1 percent for medical doctors' patients; 26.2 percent for osteopaths' patients; and 20.5 percent for chiropractors' patients. These differences held as well for the patient groups when surgery was excluded.

- **Of the patients who incurred compensable injuries, chiropractic patients were less likely to be hospitalized for treatment.**

  This finding held for claimant groups that did and did not include patients who underwent surgery. For both surgery and non-surgery groups, 23.2 percent of chiropractic patients with compensable injuries were hospitalized. For medical patients (surgery) 57.7 percent were hospitalized (non-surgery), 56.0 percent for osteopathic patients (surgery), 49.3 percent were hospitalized, and (non-surgery) 48.9 percent were hospitalized.

- **When the average number of services and average cost per service were compared, chiropractic care represented a relatively cost-effective approach to the management of work-related back injuries.**

  For the claimant group, which included surgery patients, the following findings were noteworthy: (a) For non-hospital treatment, chiropractors received the highest average (mean) payments per patient ($486), compared to osteopaths ($155) and medical doctors ($343). However, chiropractors provided a much greater average number of services per patient (29) than osteopaths (8) or medical doctors (14). Furthermore, the average amount paid per service was lowest for chiropractors ($16.31), compared to osteopaths ($18.40) and medical doctors ($23.87).

  The average cost for hospitalized patients varied significantly by provider. The average cost of hospital services for patients of medical doctors was $3,640; for patients of chiropractors $1,503; and for osteopathic patients $1,973.

  The total cost of all treatment including hospital and non-hospital services was $52 million. Of this total, medical treatment accounted for 88.7 percent ($45.5 million), or $1,100 per patient; chiropractic treatment accounted for 10.1 percent ($5.2 million), or $588 per patient; and osteopathic treatment accounted for 1.1 percent ($0.58 million), or $411 per patient.

  When surgery patients were excluded (1,695 required surgery), the comparative differences among the three doctors again held.

Appendix B

Type of Law and Insurance Requirements

Reprinted with permission from the Chamber of Commerce of the United States from the 1995 Edition of *Analysis of Workers' Compensation Laws*. *Analysis of Workers' Compensation Laws* may be ordered by calling 1-800-638-6582.

CHART I — TYPE OF LAW AND INSURANCE REQUIREMENTS

JURISDICTION	TYPE OF LAW	INSURANCE	SELF-INSURANCE	PENALTIES ON FAILURE TO INSURE
Alabama	Compulsory	Required	Individual and group	Fine of not less than $100 nor more than $1,000. Employer may be enjoined from doing business and liable to suit with defenses abrogated and double amount of compensation.
Alaska	Compulsory	Required	Permitted	Class B or C felony (up to 1 year imprisonment, $10,000 fine, or both). Board may enjoin use of labor. Employer liable to suit with defenses abrogated, and employer negligence presumed proximate cause of injury. Individuals in charge of corporation personally liable for compensation.
American Samoa	Compulsory	Required	Permitted	Misdemeanor; fine up to $1,000 or imprisonment up to 1 year, or both. Employer liable to suit with defenses abrogated.
Arizona	Compulsory	Required	Individual and group	Employer liable to suit with defenses abrogated; 10% penalty of award, expenses, and attorney's fees, or $500 (whichever is greater) plus 10% interest on amount paid from fund and penalty; award paid from Special Fund. Injunction against doing business in state.
Arkansas	Compulsory	Required	Individual and group[1]	Fine up to $10,000 or Class D Felony; employer liable to suit with defenses abrogated.
California	Compulsory	Required	Individual and group	Employer may be enjoined from doing business. Mandatory penalty upon issuance of stop order is $1,000 per employee, (maximum $100,000). If a claim is filed and an employer has not secured coverage, the employer is liable to pay $1,000 per employee in non-compensable cases and $5,000 per employee in compensable cases. Failure to obey a stop order is a misdemeanor; penalty is fine up to $1,000, imprisonment up to 60 days, or both. $50 penalty for failure to respond to Director's inquiry. Penalties are paid into the Uninsured Employers' Fund and constitute lien on employer's assets. Employee may sue for damages with employer's defenses abrogated and file for compensation. Any person who makes or causes to be made any knowingly false or fraudulent material statement or material representation for the purpose of obtaining or denying workers' compensation benefits or payments is guilty of a felony.
Colorado	Compulsory	Required	Individual and group	Compensation increased 50% or employer liable to suit with defenses abrogated (at option of employee). Employer may also be enjoined from doing business.
Connecticut	Compulsory	Required	Permitted	Fine of not more than $10,000 for failure to insure. Employer may be enjoined from entering into any contracts of employment.
Delaware	Compulsory	Required	Individual and group	Fine of 10 cents per day per employee (maximum $50, minimum $1 per day); if default continues for 30 days employer may be enjoined from doing business. Employer liable to suit with defenses abrogated.
District of Columbia	Compulsory	Required	Permitted	Civil fine of not less than $1,000 and not more than $10,000.
Florida	Compulsory	Required	Individual and group[2]	Fine of $1,000 or twice the amount the employer would have paid during periods it illegally failed to secure coverage in the preceding 3-year period, whichever amount is greatest. Failure to provide coverage is deemed an immediate and serious danger to public health, safety, or welfare sufficient to justify stop-work order issuance and $100 daily penalty until compliance is achieved. Subject to prosecution for third degree felony to knowingly fail to secure coverage if required.
Georgia	Compulsory	Required	Individual and group	Misdemeanor. Compensation may be increased 10% plus attorney's fees. Penalty up to $50 per day.
Guam	Compulsory	Required	Not permitted	Uninsured employers may be sued at law or in admiralty. Insured employer liability is exclusive for contribution among joint tort feasors against the employer.
Hawaii	Compulsory	Required	Individual and group	$250 or $10 per employee per day during default, whichever is greater. Employer may be enjoined from doing business.
Idaho	Compulsory	Required	Permitted	Misdemeanor. Employer may be liable for penalty of $2 per day per employee or minimum penalty of $25, whichever is greater, for each day failure continues. May be enjoined from doing business. Additional penalties include $500 for the second violation and $1,000 for subsequent violations.
Illinois	Compulsory	Required	Individual and group	Fine up to $500 for each day's default; employer liable to suit.
Indiana	Compulsory	Required	Permitted[3]	Class A infraction—maximum fine $10,000. Uninsured employer may be liable for medical and legal expenses plus double compensation and may be enjoined from doing business.
Iowa	Compulsory	Required	Individual and group	Employer liable to suit with defenses abrogated and presumption of negligence of employer. In coal mining, employer is liable to penalty of $10 to $100 per day and may be enjoined for further noncompliance.
Kansas	Compulsory	Required	Individual and group	Employer liable to suit with defenses abrogated.
Kentucky	Compulsory	Required	Individual and group	Failure to secure payment of compensation—claimant may claim compensation and bring action at law or in admiralty with employer's common law defenses abrogated. Employer may be enjoined from doing business. Fine of $100 to $1,000 per offense.
Louisiana	Compulsory	Required	Individual and group	Fine of $250 per employee for the first offense and $500 per employee for the second offense.
Maine	Compulsory	Required	Individual and group	Employer liable for civil penalty of up to $10,000 payable to Employment Rehabilitation Fund. Corporate employers subject to revocation or suspension of its authority to do business. Class D crime. Employer liable to suit with defenses abrogated.
Maryland	Compulsory	Required	Individual and group[4]	Fine of $500 to $5,000 and/or imprisonment for not more than 1 year. Additional penalty for failure to comply with Commission's orders amounting to 6 months insurance premiums. Employer also liable to suit with defenses abrogated. Other insurers assessed to pay unpaid claims of insolvent insurer. Fine of $300 and 15% penalty on award payable to Uninsured Employers' Fund.
Massachusetts	Compulsory	Required	Individual and group	Fine of not more than $1,500 or imprisonment for not more than 1 year, or both; employer liable to suit with defenses abrogated. Civil penalties for failure to insure include: stop-work orders, debarred from state and municipal contracts, and a $100 per day fine for each day employer operates after stop-work order.
Michigan	Compulsory	Required	Individual and group	Fine of $1,000 or imprisonment for 30 days to 6 months, or both; employer liable for damages.
Minnesota	Compulsory	Required	Individual and group	Penalty of up to $1,000 per employee per week during which the employer was not in compliance. Employer may be enjoined from further employment. Intentional noncompliance is gross misdemeanor. Employer liable to suit with some defenses abrogated.
Mississippi	Compulsory	Required	Individual and group[5]	Fine up to $1,000 or 1 year imprisonment or both; civil penalty up to $10,000. Employer also liable to suit with defenses abrogated.

Chart I — Type Of Law And Insurance Requirements

Jurisdiction	Type Of Law	Insurance	Self-Insurance	Penalties On Failure To Insure
Missouri	Compulsory	Required[8]	Individual and group	Employer liable to suit with defenses abrogated. Worker may receive medical and/or death benefits out of Second Injury Fund and employer is liable for an amount equal to twice the annual estimated premium of employer or $25,000, whichever is greater.
Montana	Compulsory	Required	Individual and group[7]	Division must enjoin uninsured employer from doing business. Double amount of unpaid premiums assessed as penalty (minimum $200). Employer liable for compensation payable up to $50,000. Employer automatically negligent if no coverage obtained. Penalties payable to Uninsured Employers' Fund.
Nebraska	Compulsory	Required	Permitted[8]	Employer liable to suit with defenses abrogated, $1,000 fine maximum, 1 year imprisonment, or both; may be enjoined from doing business.
Nevada	Compulsory[9]	Required in state fund	Individual[10]	Employer liable to suit with defenses abrogated and may be enjoined from doing business; misdemeanor punishable by a fine up to $500, administrative fines up to $10,000.
New Hampshire	Compulsory	Required	Individual and group	Penalty of $2,500 plus $100 per employee per day. Employer may be enjoined from doing business and injured worker may sue for damages. Employer shall be guilty of a misdemeanor.
New Jersey	Elective[11]	Required	Permitted	Uninsured employers are subject to fine of $1,000 and an assessment of 15% of any award not to exceed $5,000. Willful failure to provide insurance is a crime of the fourth degree. In addition, an assessment of $1,000 may be imposed for every 10-day period that insurance is not provided.
New Mexico	Compulsory	Required	Individual and group	Employer may be enjoined from doing business and/or fined up to $5,000.
New York	Compulsory	Required	Individual and group	Fine of $500 to $2,500 or imprisonment for up to 1 year, or both, with fines to $7,500 for repeated offenses. Employer liable to suit with certain special defenses abrogated. Additional fine of $250 for each 10-day period of no coverage, or a sum not to exceed 2% of payroll for period of no coverage.[12]
North Carolina	Compulsory	Required	Individual and group[13]	Misdemeanor punishable by penalty of $1.00 per day per employee (maximum $100, minimum $50 per day), imprisonment, or both. Employer liable to suit with common law defenses abrogated.
North Dakota	Compulsory	Required in state fund	Not permitted	Misdemeanor punishable by $500 fine, 1 year imprisonment, or both. Uninsured employer liable for damage or injuries or death and cannot avail himself of common law defense. Employer may be enjoined from employing uninsured workers.
Ohio	Compulsory	Required in state fund	Permitted	Minor misdemeanor—fine up to $100. If willful, second degree misdemeanor—fine up to $750, imprisonment up to 90 days, or both. Employer may be enjoined from doing business. Employer is also liable to suit with defenses abrogated.
Oklahoma	Compulsory	Required	Individual and group	Fine of $250 per employee for first offense, $500 per employee for second offense, up to $10,000. Penalty of misdemeanor or up to 6 months in jail. The Commissioner of Labor can issue cease-and-desist order against an employer who is cited for 2 offenses of failing to obtain workers' compensation insurance.
Oregon	Compulsory	Required	Individual and group	Employer is liable to suit with defenses abrogated. Enjoined from hiring workers. Liable for payment of all claims plus administrative costs. Fine up to $1,000 for first violation, to $25 per day for subsequent violations; additionional fines to $5,000 based on extent of injury.
Pennsylvania	Compulsory	Required	Permitted	Fine of $500 to $2,000 per day or not more than 1 year imprisonment, or both. Employer liable to suit with defenses abrogated.
Puerto Rico	Compulsory	Required in territorial fund	Not permitted	Misdemeanor, fine of $500 maximum, or imprisonment for not more than 6 months or both. Employer liable to suit with defenses abrogated. Penalty 30% of compensation (minimum $10.00).
Rhode Island	Compulsory	Required	Individual and group	Misdemeanor. Fine of $1,000 and/or 1 year imprisonment. Corporate officer liable personally; employer liable to suit with defenses abrogated.
South Carolina	Elective	Required[14]	Individual and group	If employer does not reject and fails to insure, fine of 10 cents per day per employee (maximum $50, minimum $1 per day); if default continues for 30 days, employer may be enjoined from doing business. Employer liable to suit with defenses abrogated. Willful failure to insure is misdemeanor punishable by fine of $100 to $1,000, or imprisonment of 30 days to 6 months, or both.
South Dakota	Compulsory	Required	Individual	Employer liable to suit for damages or double compensation and medical care as benefits.
Tennessee	Compulsory	Required	Individual and group	Administrative fine of $5,000 for every 30 days of willful refusal and noncompliance.
Texas	Elective	Required[15]	Permitted[16]	Employer liable to suit with defenses abrogated.
Utah	Compulsory	Required	Permitted	Fine of not more than 1-1/2 times the premium employer would have paid during period of non-compliance. Employer liable to suit with defenses abrogated. Costs and attorney's fees in civil suit. Employers and officers guilty of a misdemeanor. Employer liable for all compensation paid from Uninsured Employers' Fund plus interest, costs, and attorney's fees.
Vermont	Compulsory	Required	Permitted	Failure to secure compensation—fine up to $50 per day, up to a maximum of $1,500. Fine increases to $150 per day 30 days after notice by Commissioner.
Virgin Islands	Compulsory	Required in territorial fund	Not permitted	Employer liable for compensation and expenses plus penalty equal to 30% of compensation and expenses. Employer liable to suit with defenses abrogated. Fine up to $500 or imprisonment up to 6 months, or both. Interest on premiums in default. Employer may be enjoined from doing business.
Virginia	Compulsory	Required	Individual and group	Civil penalties of $500 to $5,000. Employer liable to suit with defenses abrogated and may be enjoined from doing business. Intentionally uninsured employer commits Class 2 misdemeanor.
Washington	Compulsory	Required in state fund	Permitted[17]	Employer may be enjoined from doing business. Willful failure is misdemeanor—fine is $25 to $100 daily, 50% to 100% of claim cost plus $500, or twice the unpaid premium (whichever is greater).
West Virginia	Compulsory	Required in state fund	Permitted	Employer liable to suit with defenses abrogated, all past premiums plus interest, reimbursement to state fund for claims paid on employer's behalf for failure to subscribe; may be enjoined from doing business in state.
Wisconsin	Compulsory	Required	Permitted	Forfeit $10 to $100 and double the evaded premium with a minimum payment of $750 or $100 per day for the first 7 days. Each day is a separate offense. Employer may be restrained from doing business pending compliance. Employer liable for all benefits awarded on uninsured claims.
Wyoming	Compulsory	Required in state fund[18]	Not permitted	Fine of not more than $750 for first conviction; fine of not more than $10,000 for second conviction or subsequent conviction, plus .02% interest per month on unpaid balance. Employer may be enjoined from doing business and liable to suit with defenses abrogated.
F.E.C.A.	Compulsory	Federal appropriation	Individual and group	
Longshore Act	Compulsory	Required	Not permitted	Fine of not more than $10,000 or 1 year in prison or both for failure to secure payment of compensation.

CHART I — TYPE OF LAW AND INSURANCE REQUIREMENTS

JURISDICTION	TYPE OF LAW	INSURANCE	SELF-INSURANCE	PENALTIES ON FAILURE TO INSURE
Alberta	Compulsory	Required in provincial fund	Not permitted	Failure to furnish required security—Board may order employer to cease employment; violation—fine up to $200 daily. Failure to submit statement of wages—up to 15% assessment plus penalty up to half of compensation payable, maximum $500. Failure to pay assessment—up to 24% penalty per annum. Board certificate has same effect as court judgement; employer's goods may be seized.
British Columbia	Compulsory	Required in provincial fund	Not permitted	Failure to submit statement of wages—compensation payable plus percentage of assessment set by Board. Failure to pay assessment—unpaid amount plus cost of collection and percentage penalty, compensation payable, and employer may be enjoined from operating.[19]
Manitoba	Compulsory	Required in provincial fund	Permitted for Crown	Penalty for late filing of payroll return is 5% of employer's assessment; penalty for failure to submit return is 10% of employer's assessment. Inaccurate reporting may result in a penalty of 15% of the difference between the amount reported by the employer and the amount determined by the Board. In all cases, maximum penalty is $5,000.
New Brunswick	Compulsory	Required in provincial fund[20]	Permitted for Crown	Failure to submit statement of wages—maximum penalty $500 plus percentage fixed by Board. Failure to pay assessment —employer liable for compensation payable plus percentage penalty and costs of collection.
Newfoundland	Compulsory	Required in provincial fund	Not Permitted	Failure to submit statement of wages—maximum penalty $2,000. Failure to pay assessment—assessment plus costs of collection and a percentage penalty. Employer may be held liable for costs of claims or injuries during period of default.
Northwest Territories	Compulsory	Required in territorial fund	Permitted	Failure to submit statement of wages or to pay assessment—employer liable for compensation payable and assessment plus percentage of assessment as penalty established by Board or regulations.
Nova Scotia	Compulsory	Required in provincial fund	Not Permitted[21]	Employer liable for compensation payable. Failure to submit statement of wages—employer liable for unpaid amount and costs of collection plus 5% penalty; 1% penalty for each month in default; $50 if annual statement. Failure to pay assessment—2% penalty applied on the premium for the first month plus 1% for each subsequent month in default.
Ontario	Compulsory	Required in provincial fund	Permitted[22]	Failure to submit statement of wages upon commencing operations and at other required times—penalty plus liability for additional percentage of assessment and costs of claim at discretion of Board.
Prince Edward Island	Compulsory	Required in provincial fund	Not Permitted	Failure to report payroll or pay assessment—penalty of $100 per week of default plus 2% of amount unpaid after 1 month and 1% for each additional month. Failure to pay assessment—employer may be prevented from operating.
Québec	Compulsory	Required in provincial fund	Not Permitted[23]	Failure to submit statement of wages upon commencing operations or at other required times—penalty of 5% of the assessment unpaid for the first month plus the interest on the assessment for subsequent months, in addition to a maximum fine of $1,000. Failure to pay assessment, employer liable for 10% of claim which cannot be less than $100.
Saskatchewan	Compulsory	Required in provincial fund	Not Permitted	Failure to submit statements of wages upon commencing operations or at other required times—maximum fine $1,000. Failure to pay assessment, employer liable for percentage penalty to be established by Board and may be enjoined from operating.
Yukon Territory	Compulsory	Required in territorial fund	Not Permitted	Upon failure to submit statement of wages upon commencing operations or at other required times and failure to pay assessment, an employer is liable for a percentage penalty established by the Board, and employer may be enjoined from operating.
Canadian Merchant Seaman's Act	Compulsory	Required	At discretion of Board	Failure to insure or cover by other means satisfactory to the Board may cause ship to be detained by Customs.

[1] **Arkansas** — Municipalities with more than 70,000 population may self-insure on individual or group basis.
[2] **Florida** — Application for workers' compensation coverage under a group self-insurance fund must contain the following statement: "This is a fully assessable policy. If the fund is unable to pay its obligations, policyholders must contribute on a pro rata earned premium basis the money necessary to meet any unfilled obligations."
[3] **Indiana** — Except as to state and political subdivisions, banks, trust companies, and savings and loan associations.
[4] **Maryland** — Eligibility for group self-insurance is limited to counties, municipalities, and certain private employers.
[5] **Mississippi** — All self-insurers must be members of the Mississippi Workers' Compensation Self-Insurer Guaranty Association.
[6] **Missouri** — Employer engaged in mining must insure only to the extent of maximum liability for 10 deaths in any one accident.
[7] **Montana** — Private employers and public entities, other than state agencies, may establish individual or group self-insurance funds.
[8] **Nebraska** — Group self-insurance permitted for any two or more public agencies.
[9] **Nevada** — Employee temporarily working in state must prove coverage acceptable to Nevada from another state before working in Nevada.
[10] **Nevada** — Group self-insurance will be permitted 7/1/95.
[11] **New Jersey** — Statutory presumption of compulsory inclusion in every contract of hire since July 4, 1911. To elect not to be covered by the Workers' Compensation Act requires an unambiguous written agreement between employee and employer 60 days prior to the happening of an accident.
[12] **New York** — President, secretary and treasurer of a corporation are personally liable for penalties. Corporate officer who failed to obtain insurance is ineligible for benefits out of Uninsured Employers' Fund for self, surviving spouse, or dependents.
[13] **North Carolina** — All individual and group self-insurers must be members of North Carolina Self-Insurance Guaranty Association as a condition of authority to self-insure.
[14] **South Carolina** — Unless employer rejects.
[15] **Texas** — If employer accepts.
[16] **Texas** — Except for state and political subdivisions. Beginning 1/1/93, self-insurance is permitted upon Commission's approval of each inquiry.
[17] **Washington** — Group self-insurance permitted for school districts and hospitals.
[18] **Wyoming** — For "extra-hazardous" industries and occupations only.
[19] **British Columbia** — Employer directly liable for compensation if injury was caused by employer's gross negligence or lack of an accident prevention program; maximum $31,910.57.
[20] **New Brunswick** — Subject to approval of Lt.-Gov. in Council, Board may make arrangement for insurance or reinsurance.
[21] **Nova Scotia** — Individuals whose industries fall outside the scope of the WCB of Nova Scotia Act may elect to pay for compensation insurance under the Board's Special Coverage plan.
[22] **Ontario** —Employers defined as Schedule 2 employers under the Ontario Workers' Compensation Act are permitted to self-insure.
[23] **Québec** — Permitted only for employer operating an interprovincial or international railway or maritime transport firm.

Appendix C

Waiting Period for Income/ Medical Benefits

Reprinted with permission from the Chamber of Commerce of the United States from the 1995 Edition of *Analysis of Workers' Compensation Laws*. *Analysis of Workers' Compensation Laws* may be ordered by calling 1-800-638-6582.

Chart IX — Waiting Period For Income/Medical Benefits

Jurisdiction	Waiting Period[1]	Retro-active Period	Unlimited[3]	Choice Of Physician[2]: Employer	Choice Of Physician[2]: Employee	Medical Benefits: Artificial Appliances Furnished	Special Provisions
Alabama	3 days[4]	3 weeks	Yes	initial choice	[5]	Yes	Employer must replace appliances damaged in work-related accident and provide physical and vocational rehabilitation.
Alaska	3 days	4 weeks	Yes	Yes	Yes	Yes	Injury includes damages to eyeglasses, dentures, hearing aids or any prosthetic devices.
American Samoa	3 days	2 weeks	Yes			Yes	
Arizona	7 days	2 weeks	Yes		Yes	Yes	Prayer or spiritual treatment by agreement.
Arkansas	7 days	2 weeks	Yes[6]	Agency may change	Agency may change	Yes	Spiritual treatment by agreement. Eyeglasses, contact lenses, and hearing aids.
California	3 days[7]	14 days[7]	Yes		If employee has not redesignated a physician before time of injury, they may change to treating physician of their choice 30 days after injury. If employer offers managed care options, employer control of care may extend up to 365 days.	Yes	Includes x-ray reports, medical reports, and testimony and laboratory fees reasonably required to prove a claim.[8] If requested, an employee may change physicians at any time, 1 time only.
Colorado	3 days	2 weeks	Yes	Initial choice[9]		Yes	Employer must repair/replace appliances damaged in compensable injury.
Connecticut	3 days	1 week	Yes		From state list	Yes[10]	Worker compensated for time lost due to medical attention. Employer must repair/replace appliances damaged in employment, including eyeglasses, contact lenses, hearing aids, or dentures where injury to face or head. Prayer or spiritual treatment with Commissioner's approval.
Delaware	3 days[11]	7 days[11]	Yes		Yes	Yes	Employer must replace prosthesis and furnish hearing aids as needed.
District of Columbia	3 days	2 weeks	Yes		Yes	Yes	Spiritual treatment by agreement. Medical care furnished or scheduled to be furnished, is subject to utilization review for determinations of necessity, character or sufficiency of care or service.
Florida	7 days	14 days	Yes[12]	Yes		Yes	Injury includes damage to dentures, eyeglasses, and prosthetic devices in conjunction with accident. Employer must provide custodial care.
Georgia	7 days	21 days	Yes	Agency may change	From employer list	Yes[13]	Employee may be examined (at employer's expense) by physician of choice within 60 days of receiving any income benefits.
Guam	3 days	14 days	Yes		After application to Commissioner for change	Yes	
Hawaii	3 days		Yes		Yes	Yes	Maximum monthly attendant's allowance—4 times the SAWW
Idaho	5 days[14]	2 weeks[14]	Yes	Yes		Yes	Christian Science treatment permitted.
Illinois	3 days[15]	2 weeks	Yes		Yes	Yes	Spiritual treatment by agreement. Employer must repair/replace appliances damaged in compensable accident.
Indiana	7 days	3 weeks	Yes	Yes		Yes	Employer pays reasonable expenses, including travel, food and lodging, for treatment outside county. Spiritual treatment by agreement. Employer must replace artificial members when medically required.
Iowa	3 days[16]	14 days	Yes	Yes		Yes	Employer must repair/replace appliances.
Kansas	7 days	3 weeks	Yes	Yes	[17]	Yes	Prayer or spiritual treatment permitted by agreement.[18]
Kentucky	7 days	2 weeks	Yes		Yes[19]	Yes	
Louisiana	7 days	6 weeks	Yes		Yes	Yes	Employer must repair/replace appliances.
Maine	7 days[20]	2 weeks	Yes[21]	Yes[22]		Yes	Chiropractic services authorized. Prayer or spiritual means of treatment by an accredited practitioner.
Maryland	3 days	2 weeks	Yes	Yes		Yes	Employer must repair/replace appliances. If employer or insurer fails to pay for treatment or services within 45 days, Commission may assess a fine of more than 20% of approved charges. Also, subject to interest.
Massachusetts	5 days	21 days	Yes		Yes	Yes	
Michigan	7 days	2 weeks	Yes	Initial choice[23]		Yes	
Minnesota	3 days	10 days	Yes	Agency may change	Yes[24]	Yes	Christian Science treatment by agreement. Chiropractic and podiatric treatment authorized.
Mississippi	5 days	2 weeks	Yes		Yes	Yes	Employee's choice limited to one (1) provider.
Missouri	3 days	2 weeks	Yes	Agency may change		Yes	Prayer or spiritual treatment by agreement. Chiropractic care authorized.
Montana	6 days, 48 hours[25]		Yes		Yes	Yes	Reasonable lifetime medical and travel benefits provided. Limitations apply once claimant reaches maximum stability. Certain co-payments may apply. Other limitations include secondary medical services, drug reimbursement, palliative or maintenance care and unscientific treatment. Medical benefits terminated if not used for 60 consecutive months.
Nebraska	7 days	6 weeks	Yes			Yes	Employer must replace appliances damaged due to compensable injury. Employer must provide plastic surgery for disfigurement.
Nevada	5 days[26]	5 days[26]	Yes	[27]	Yes[28]	Yes	Spiritual treatment permitted.
New Hampshire	3 days	14 days	Yes[29]	Yes[30]	Yes[31]	Yes	Commissioner will adopt a medical fee schedule by 1/1/95.
New Jersey	7 days	8 days	Yes[32]	Yes		Yes	Hospital care must be semi-private, if available.

Chart IX — Waiting Period For Income/Medical Benefits

Jurisdiction	Waiting Period[1]	Retro-active Period	Unlimited[3]	Medical Benefits: Choice Of Physician[2]: Employer	Medical Benefits: Choice Of Physician[2]: Employee	Medical Benefits: Artificial Appliances Furnished	Special Provisions
New Mexico	7 days	4 weeks	Yes	Yes		Initial choice[33]	Claimant may not refuse treatment reasonably essential to provide recovery.[34]
New York	7 days	2 weeks	Yes		From state list	Yes[35]	Employer liable for x-rays, special diagnostic tests, consultations, and providing prosthetic devices. Chiropractic, dental, psychological, podiatric, nurse and midwife care may all be authorized.
North Carolina	7 days	3 weeks	Yes	Yes	Agency may change	Yes	Employer must repair/replace appliances damaged in compensable accident. Medical care includes rehabilitation services.
North Dakota	5 days[36]	5 days[36]	Yes		Yes	Yes	Employee has first choice of treating doctor; however, Bureau can redirect to different doctor through managed care program.
Ohio	7 days	2 weeks	Yes		Yes	Yes	Includes hospitalization and damages to eyeglasses, dentures, hearing aids or prostheses.
Oklahoma	3 days		Yes	Court may change	Yes	Yes	Employer must repair/replace appliances. Special provisions for hernias.
Oregon	3 days	2 weeks	Yes[37]		Yes[38]	Yes	Spiritual treatment by agreement. Employer must pay for medical exam and related services. Insurer/employer may provide medical treatment through an approved Managed Care Organization.
Pennsylvania	7 days	2 weeks	Yes	Initial choice[39]		Yes	
Puerto Rico	[40]	10 days	Yes[41]	State Agency		Yes[42]	When permanently and totally disabled, Administrator will provide orthopedic girdle, crutches, cane, wheelchair, hospital bed, and other equipment necessary.
Rhode Island	3 days		Yes[43]	Yes[44]	Yes[44]	Yes	Injury includes damage to and cost of replacement of eyeglasses, hearing aids and prosthetic devices.
South Carolina	7 days	2 weeks	Yes	Yes		Yes	Injury includes damage to and cost of replacement of eyeglasses, hearing aids and prosthotic devices.
South Dakota	7 days[45]	7days	Yes		Yes	Yes	Employer must repair/replace appliances damaged in compensable accident.
Tennessee	7 days	2 weeks	Yes		From employer list	Yes	Provides for nursing services and treatment by chiropractors.
Texas	7 days	4 weeks	Yes[46]		Yes[47]	Yes	
Utah	3 days	2 weeks	Yes	Agency may change[48]		Yes	Reasonable amounts awarded for repair/replacement of artificial appliances.
Vermont	3 days[49]	10 days	Yes	Agency may change		Yes	Injury includes damage to and cost of replacement of eyeglasses, hearing aids and prosthetic devices. Employer must not withhold any wages from an employee if the employee's absence is for an examination or treatment of a work injury.
Virgin Islands	1 day	1 day	$40,000		Yes, agency may change	Yes	$75,000 limit for specialized treatment not available in the Virgin Islands.[50]
Virginia	7 days	3 weeks	Yes		From employer list	Yes	Employer must repair/replace appliances damaged in compensable accident. Employer may be ordered to furnish wheelchair and make alterations to home, maximum $25,000.
Washington	3 days	2 weeks	Yes		Yes	Yes	Employees pay half of medical aid premiums. Department will repair/replace appliances damaged in compensable accident.
West Virginia	3 days	1 week	Yes[51]		Yes	Yes	Payment for prosthetic/orthotic appliances will not be made until appliance is deemed serviceable. Repair/replacement of glasses damaged in an accident not paid for unless there was compensable injury.
Wisconsin	3 days	1 week	Yes		Yes	Yes	Repair/replacement of appliance is limited to normal wear and tear. Dental care, eyeglasses and hearing aids are also covered. Repair/replacement of eyeglasses and hearing aids not paid for unless damaged in compensable accident. Employer or insurer may request that a physician examine injured employee to determine reasonableness of claim.
Wyoming	3 days	8 days	Yes	Second opinion	Yes	Yes	All medical bills audited according to promulgated fee schedule.
F.E.C.A.	3 days[52]	14 days	Yes		Yes	Yes	Additional amount (up to $1,500) monthly for medical attendant.
Longshore Act	3 days	2 weeks	Yes		Labor Secretary may change	Yes	Consent for specialist, if needed. Spiritual treatment permitted.
Alberta	1 day	1 day	Yes		Yes	Yes	Appliances repaired/replaced by Board. Board may repair/replace garment damaged in compensable accident. Clothing allowance for wear due to prosthetic or wheel-chair—up to $200 per year for upper body and $300 per year for lower body. Attendance allowance—determined by the degree of personal care required and will reflect actual and reasonable costs or competitive rates.

Chart IX — Waiting Period For Income/Medical Benefits

Jurisdiction	Waiting Period[1]	Retro-active Period	Medical Benefits: Unlimited[3]	Choice of Physician[2]: Employer	Choice of Physician[2]: Employee	Artificial Appliances Furnished	Special Provisions
British Columbia	1 day		Yes		Yes	Yes	Appliances repaired/replaced at Board's discretion, regardless of personal injury (includes eyeglasses, dentures and hearing aids unless worker is at fault). Board may provide eyeglasses for serious visual impairment caused by work injury. Personal care allowance—maximum $44 per day. Annual clothing allowance for wear due to prosthesis—$287.50 for upper limb, $415.50 for lower limb, $578 for both.
Manitoba	1 day		Yes		Yes	Yes	Medical aid includes repair/replacement of damaged or lost clothing or personal appliance if personal injury. Personal appliances include eyeglasses, contact lenses, and prosthetic devices. Clothing allowance for additional wear due to prosthetic device —$186 for upper extremity, $373 for lower extremity, $373 for orthotic device. Attendance allowance —$175 to $1,215 (monthly amounts). Emergency expenditures —$5,000 maximum.
New Brunswick	3 days	1 day	Yes		Yes	Yes	Medical aid includes repair/replacement of appliances. Clothing allowance for wear due to prosthesis is up to $428 annually. Care allowances from $215.92 to $719.73 monthly.
Newfoundland	1 day	[53]	Yes	No	Initially yes; should consult Commission before changing	Yes	Commission may repair/replace appliances and may pay daily allowance for treatment away from home. Clothing allowance for wear due to prosthetic device—$150 to $500 annually. Personal care allowance—$100 to $1,250 monthly.
Northwest Territories	1 day	1 day	Yes		Initial choice	Yes	Clothing allowance for wear due to prosthetic device—$100. Board may repair/replace appliances damaged in a compensable accident.
Nova Scotia	1 day[54]	[55]	Yes	Yes[56]	Yes	Yes	Attendant's allowance—$300 monthly. Clothing allowance for wear due to prosthetic device—$350 per year. Board may repair/replace appliances, and renew eyeglasses (replaced if damage in work-related accident).
Ontario	1 day	[57]	Yes		Initial choice	Yes	No maximum attendance allowance; payment will be actual cost. Clothing allowance for wear due to prosthetic device—$252.77 for upper extremity and $505.55 for lower extremity.[58]
Prince Edward Island	1 day	1 day	Yes		Initial choice	Yes	Appliances repaired/replaced at Board's discretion. Clothing allowance for wear due to prosthetic device, for para- and quadriplegics is $214 annually.
Québec	Day of injury[59]		Yes		Yes	Yes	Commission may repair/replace prosthesis damaged in course of employment. Additional clothing allowance for wear due to prosthetic device—$428. Attendance allowance—$50 to $1,140.
Saskatchewan	[60]	[60]	Yes		Yes	Yes	Clothing allowance for wear due to prosthetic device—$195 for arm, $430[61] for leg. Personal care allowance—$279 to $1,442.[62]
Yukon Territory	1 day	1 day	Yes		Yes	Yes	Clothing allowance—$149 for upper limb and $303 for lower limb. Personal care allowance—$7 to $24 per day, independence allowance $100 per month.
Canadian Merchant Seaman's Act	3 days	3 days	Yes		Yes	Yes	Employer must keep appliances in repair or replace, at Board's discretion.

[1] If disability continues for longer than stated periods, compensation is paid for the waiting period. Waiting periods do not apply to medical care, which is furnished from the first day of injury.
[2] Information supplied by Division of State Workers' Compensation Programs, U.S. Department of Labor.
[3] Many states have introduced medical fee schedules to control payments made to providers for medical care under the workers' compensation programs. Medical benefits remain "unlimited" in those states indicated, but employees and insurers are only liable for the amounts set forth in the medical fee schedule. Employees cannot be held accountable for any amount charged above and beyond the medical fee schedule.
[4] **Alabama** —Temporary disability only.
[5] **Alabama** — Employee can select second physician from a panel selected by employer. Employer or employee may select initial physician for physical rehabilitation.
[6] **Arkansas** — Employer liability may cease 6 months after injury where no time is lost from work, or 6 months after claimant returns to work, or maximum $10,000 has been paid, unless the employer waives rights or the Commission extends time and dollar limits.
[7] **California** — Waiting period also terminated by hospitalization.
[8] **California** — Psychologists included within definition of physician and treatment permitted. Personal chiropractor allowed if employee has previously notified employer that chiropractic treatments were being rendered.
[9] **Colorado** — Director maintains a list of physicians who serve as a medical review panel and that will perform independent medical exams upon any request by a party to a workers' compensation proceeding.
[10] **Connecticut** — By court decision, *Olmstead v. Lamphier*, 93 Conn. 20, 104 A. 488 (1918).
[11] **Delaware** — No waiting period if incapacity results in hospitalization or is caused by amputation of member.
[12] **Florida** — Expenses subject to medical fee schedule; carrier or employee is allowed to select an independent medical examiner in disputes concerning overutilization, medical benefits, compensability, or disability.
[13] **Georgia** — Included in total amount allowed for medical care.
[14] **Idaho** — Waiting period also terminated by hospitalization.
[15] **Illinois** — TTD only.
[16] **Iowa** — No waiting period for PP disability.
[17] **Kansas** — Employee may consult a physician of his own choice at any time but employer is only liable for fees and charges up to $500 for treatment. This unauthorized medical treatment cannot be used to obtain functional impairment ratings.
[18] **Kansas** — Director must adopt rules which establish a maximum medical fee schedule, subject to approval of an advisory panel. Fee schedule to be reviewed annually.
[19] **Kentucky** — Employee must choose a "gatekeeper" physician to coordinate medical care. Employee may change his gatekeeper physician once without authorization from the medical payment obligor. Gatekeeper physician must file a treatment plan with medical payment obligor or employer.
[20] **Maine** — Fire fighters are exempt from waiting period and receive compensation from date of disability.
[21] **Maine** — Subject to a medical fee schedule.
[22] **Maine** — Employer has initial choice of physician. After 10 days, employee may select different health care provider. Employee may not change health care provider more than once without employer or Board approval. This does not apply if referred to specialist.
[23] **Michigan** — Employer may choose physician for first 10 days of treatment, thereafter, the employee may choose own physician by giving notice to employer.
[24] **Minnesota** — If the employer has a managed care plan, the employee must choose among the physicians within the plan, with certain exceptions.
[25] **Montana** — Waiting period refers to number of days (hours) on which a worker has lost wages and is totally disabled and unable to work.
[26] **Nevada** — Waiting and retroactive period is at least 5 consecutive or cumulative days within a 20-day period.
[27] **Nevada** — Employers in Las Vegas and Reno choose Managed Care Organization.
[28] **Nevada** — In Las Vegas and Reno, employee chooses treating physician within Managed Care Organization selected by employer. In remaining counties, employees have choice of any treating physician.
[29] **New Hampshire** — Unlimited for those not under a managed care program.
[30] **New Hampshire** — If approved managed care program, employee must choose from network.
[31] **New Hampshire** — Employee must choose within managed care network.
[32] **New Jersey** — Employer responsible for all necessary and reasonable costs of medical care, including chiropractic authorized and selected by employer. Employee has no freedom of choice.
[33] **New Mexico** — Employer chosen physician treats injured worker for first 60 days. Employer, however, may allow employees to select own physician instead.
[34] **New Mexico** — By court decision, *Brooks v. Employers National Insurance Co.*, 688 P.2d25 (1984). A schedule of maximum allowable payments for non-hospital medical services was adopted 1/1/92. A cost-to-charge ratio for payment of hospital charges established 4/1/91.
[35] **New York** — Managed care available to employees of employers participating in pilot project for cost containment.
[36] **North Dakota** — Only scheduled work days lost in the first 5 days are paid if disability period is 14 days or less; all 5 days are paid if disability is 15 days or more.
[37] **Oregon** — Costs subject to medical fee schedule, as determined by Director. Treatment may be restricted to MCO plan. Limitations on palliative care.
[38] **Oregon** — May choose physician within state. Allowed 2 changes; changes thereafter require Director's approval. If governed by MCO, treatment and choice may be restricted to plan.
[39] **Pennsylvania** — Only if a list of at least 6 designated health care providers. At least 3 of the providers must be physicians; no more than 2 of the provider may be coordinated care organizations or combinations thereof is posted and for the first 30 days after the first visit.
[40] **Puerto Rico** — Compensation paid from day of first medical treatment.
[41] **Puerto Rico** — Limitations on palliative treatments. *Sergio Torres Garcia v. State Insurance Fund*, June 30, 1981, P.R. Supreme Court.
[42] **Puerto Rico** — Only in cases of PT disability.
[43] **Rhode Island** — Non-hospital costs subject to medical fee schedule, to be updated annually, as determined by the Director in consultation with selected others.
[44] **Rhode Island** — If employer has a Preferred Provider Network (PPN) in place, employee has choice of initial physician, but subsequent choices may be made from the PPN.
[45] **South Dakota** — Consecutive days.
[46] **Texas** — Medical charges subject to medical fee guidelines.
[47] **Texas** — Effective 1/1/93, employee will choose physician from a commission-approved list and will be allowed one change of physician.
[48] **Utah** — Employee is allowed one change of physician.
[49] **Vermont** — Total disability pay.
[50] **Virgin Islands**—Total amount of $75,000 includes travel and accommodations.
[51] **West Virginia** — Costs subject to medical fee schedule, updated periodically.
[52] **F.E.C.A.** — Waiting period begins running after 45 days continuation of pay.
[53] **Newfoundland** — Employer payment for day on which accident occurs.
[54] **Nova Scotia** — No waiting period for permanent partial disability.
[55] **Nova Scotia** — Employer required to pay first accident day, Board pays thereafter.
[56] **Nova Scotia** — If so required by employer.
[57] **Ontario** — Employer must pay wages and benefits for day of injury.
[58] **Ontario** — Compensation payable during disability caused by damage to prosthetic device.
[59] **Québec** — Employer must pay wages and benefits for day of injury.
[60] **Saskatchewan** — Compensation paid for the day on which the accident occurs.
[61] **Saskatchewan** — 1995 figures.
[62] **Saskatchewan** — 1994 figures.

Appendix D

Rehabilitation of Disabled Workers

Reprinted with permission from the Chamber of Commerce of the United States from the 1995 Edition of *Analysis of Workers' Compensation Laws*. *Analysis of Workers' Compensation Laws* may be ordered by calling 1-800-638-6582.

Chart X — Rehabilitation Of Disabled Workers

Jurisdiction	Source Of Fund	Maintenance Allowance	Special Provisions
Alabama	No fund established.	Board, lodging and travel, if away from home.	Physical and vocational rehabilitation to restore employee to gainful employment at furnished employer's expense. Employee's refusal results in loss of compensation for the period of refusal.
Alaska	No fund established.	Board, lodging, travel and temporary disability benefits.[1]	Employer pays full cost, on an expense incurred basis, not to exceed $10,000. Benefits may not extend past 2 years from date of plan approval. Reemployment benefits suspended for unreasonable failure to participate in approved or agreed plan.
American Samoa	Second Injury Fund.	$10 weekly paid by employer plus maintenance from Special Fund.	Commission arranges for vocational rehabilitation of permanently disabled workers.
Arizona	Special Fund tax up to 1.5% on premiums written.	Commission may authorize additional necessary awards to persons undergoing vocational rehabilitation.	Vocational rehabilitation trainees considered an employee at $200 monthly wage rate for compensation benefits.
Arkansas	No fund established.	Reasonable expenses for maintenance, travel and other necessary costs, 72 weeks maximum.	Must apply to Commission. Commission may authorize vocational rehabilitation if reasonable in relation to disability, but worker may refuse.
California	No fund established.	During rehabilitation, necessary living expenses plus either a maintenance allowance not exceeding $246 per week (which may be supplemented up to temporary disability maximum if medical condition has become permanent and stationary) or temporary disability indemnity.	Rehabilitation unit in Office of Benefit Determination. Rehabilitation program is compulsory on part of employer or carrier. Rehabilitation trainee is considered employee of training employer for insurance purposes. Employees restricted to 1 rehabilitation plan, which must be completed within 18 months. A $16,000 expenditure cap is established on all vocational rehabilitation plans. The cap applies to all expenses, counseling fees, training, maintenance allowance, and costs associated with or arising out of vocational rehabilitation services. Costs of counselor fees are capped at $4,500. Rehabilitation counselors and insurers are prohibited from referring injured worker to evaluation, training, or other rehabilitation entities in which they have a financial interest.
Colorado	No fund established for vocational rehabilitation.	TTD and plan expenses paid during plan by carrier or employee if offered and accepted.	Voluntary benefit offered by carrier after 7/2/87. If employee refuses offer of vocational rehabilitation, PT benefits will not be awarded.
Connecticut	Funded out of unified 4% assessment.	Weekly subsistence allowance during vocational rehabilitation.	Employer pays full cost of medical rehabilitation, which continues until employee reaches maximum improvement. Vocational rehabilitation is furnished by Bureau of Workers' Compensation in the Department of Human Resources.
Delaware	No fund established.	Reasonable board, lodging and travel.	Physical and vocational rehabilitation furnished at employer's expense. Employee's refusal results in loss of compensation.
District of Columbia	No fund established.	Not exceeding $50 per week.	Employer must provide vocational rehabilitation. Benefits forfeited if worker fails to cooperate.
Florida	Payments from Special Fund and assessments upon insurers and self-insurers.	Reasonable board, lodging and travel, if away from home.	Insurance carrier is required to provide reemployment services for dates of accident 9/30/89 or before and may voluntarily provide services for dates of accidents after 10/1/89. The Division of Workers' Compensation shall pay for reemployment services for dates of accident 10/1/89 and after if such services are necessary for the injured employee to return to work. Carriers shall pay at least 26 weeks of the 52 weeks of rehabilitation TT benefits that an injured employee may be entitled.[2]
Georgia	No fund established.	Reasonable board, lodging and travel, if away from home.	Rehabilitation benefits are compulsory only in catastrophic cases. Employee's unreasonable refusal may result in suspension of compensation.
Guam	State fund (appropriation).	$50 per week in addition to other compensation.	Commission directs the vocational rehabilitation of permanently disabled employees and arranges with the appropriate public or private agencies for such education.
Hawaii	No fund established.	Board, lodging, travel, tuition, books and basic materials in addition to compensation.	Rehabilitation unit within the Department of Labor and Industrial Relations makes recommendation for physical or vocational rehabilitation. Director approves services and reviews progress.
Idaho	Part of 2.5% Premium Tax Industrial Administration Fund.	Reasonable expenses for maintenance and travel.	Rehabilitation Division administers. TT or TP benefits may be payable where retraining is authorized. Period of retraining not to exceed 52 weeks unless the Commission, following application and hearing, extends the period.
Illinois	No fund established.	Maintenance costs and incidental expenses.	Physical, mental and vocational rehabilitation as necessary. Institutional care, if required.
Indiana	No fund established.		Vocational rehabilitation available to any employee unable to attain gainful employment, due to an occupational disease or injury.[3]
Iowa	No fund established.	$20 weekly in addition to other compensation for 13 weeks.	May be extended additional 13 weeks. Medical care includes physical rehabilitation.
Kansas	No fund established.	Employer must pay reasonable board, lodging and travel up to $3,500 for a 36-week period and may be required to pay up to $2,000 more, if necessary.	Employer or carrier must refer certain employees to an approved rehabilitation vendor for vocational evaluation. Employer must provide up to 36 weeks (may be extended an additional 36 weeks). Disability payments suspended for worker's refusal to participate in rehabilitation. After 90 days refusal, Director may reduce compensation to not less than PP disability payments. Post 7/1/93 accidents, any vocational rehabilitation services unders the workers' compensation act must be voluntarily provided by the employer or carrier.
Kentucky	No fund established.	Board, lodging, and travel, if away from home.	Unlimited medical rehabilitation; vocational rehabilitation up to 52 weeks (may be extended). Employee's refusal results in loss of 50% of compensation.
Louisiana	No fund established.	Board, lodging and travel paid by employer or carrier.	Employer or carrier provides up to 26 weeks of vocational rehabilitation, extendable another 26 weeks. Benefits reduced 50% for refusal of necessary rehabilitation.
Maine	Employment Rehabilitation Fund.[4]	Transportation or any extra and necessary expenses, upon board order.	Employee entitled to rehabilitation services when unable to perform work for which the employee has previous training or experience. Training, treatment or service is only 52 weeks unless board extends.[5]
Maryland	No fund established.	Up to $40 weekly paid by employer.	Workers Compensation Commission investigates claims and reports of disabilities for referral to State Department of Vocation Rehabilitation or to a private vendor. Employee is entitled to rehabilitation services with TT benefits. Employee entitled to 24 months of vocational rehabilitation. Employee's unreasonable refusal results in loss of compensation. Employer pays compensation for TT disability plus expenses of vocational assessment and rehabilitation.
Massachusetts	Paid the same as compensation by employer or insurer.[6]	Office of Education and Vocational Rehabilitation may approve room, board and travel expenses for 52 weeks.	Necessary cost of rehabilitation subject to approval by Office of Education and Vocational Rehabilitation. Benefits suspended for refusal to participate.
Michigan	No fund established.	Transportation and other necessary expenses during 52 weeks training.	Medical and vocational rehabilitation services under Workers' Compensation Bureau approved facility. Bureau may extend training period additional 52 weeks, maximum total 104 weeks.

CHART X — REHABILITATION OF DISABLED WORKERS

JURISDICTION	SOURCE OF FUND	MAINTENANCE ALLOWANCE	SPECIAL PROVISIONS
Minnesota	No fund established.	Necessary expenses, including tuition, books, travel, board, lodging and custodial daycare. Employee receives temporary total during retraining up to 156 weeks.	Qualified injured worker entitled to rehabilitation provided by/at expense of employer. If approved plan, employer to provide retraining up to 156 weeks.[7] Participant may request 25% benefit increase and is eligible for one-time relocation allowance. Employer may seek termination or suspension of benefits if worker fails to cooperate with plan. Rehabilitation not mandatory but may be requested.[8]
Mississippi	No fund established.	Up to $10 per week up to 52 weeks.	Commission cooperates with federal and state agencies.
Missouri	Partial source, Second Injury—responsible for partial payment of $40 per week, not to exceed 20 weeks, with all remaining costs to be at the expense of the employer or insurer.	Reasonable board, lodging and travel if away from home, paid by employer or insurer. TT or TP benefits paid throughout duration of rehabilitation program.	Administered by Director of Workers' Compensation Division. Division may order employer to provide transportation. Employer must pay lost wages for up to 20 weeks during physical rehabilitation. Initial rehabilitation plan may not exceed 26 weeks; employers may extend plan up to 26 weeks.[9]
Montana	Rehabilitation Fund by 1% tax upon compensation paid by insurers, self-insured, and state fund.	Disabled worker, unable to return to job at time of the injury, is eligible for up to 104 weeks of benefits at the TT rate while implementing a rehabilitation plan which can be job replacement, on-the-job training, education, training or specialized job modification. Up to 10 weeks of benefits are available while the worker is awaiting the opportunity to implement a rehabilitation plan, and 8 weeks of benefits are available for use in obtaining employment when the plan calls for job placement. Auxiliary benefits up to $4,000 available for travel and relocation expenses, as well as, implementing a rehabilitation plan.	Administered by Department of Labor and Industry rehabilitation panel. Services provided by private certified rehabilitation counselors. Employee also may be referred to the Department of Social and Rehabilitative Services. Employee's refusal may result in loss of compensation. Disputes over the rehabilitation plan resolved by the Workers' Compensation Court.
Nebraska	Vocational Rehabilitation Fund by 2% of benefits paid by carriers and self-insurers in the state in prior year.	Board, lodging, travel, tuition, fees, and books paid by fund. Temporary indemnity paid by self-insurer or carrier.	Payments into fund suspended when fund reaches $1,500,000. Assessment (2%) when fund reduced to $800,000.
Nevada	State Insurance Fund and self-insurance.	Insurer may allow maintenance as needed.	Insurer is authorized to provide necessary rehabilitation services. Employee's refusal results in loss of all benefits.
New Hampshire	No fund established.	Board, lodging, travel, books, and basic materials in addition to compensation.	Insurer must furnish rehabilitation services voluntarily, or may be ordered to do so, for 1 year more if unusual circumstances. Vocational Rehabilitation Coordinator can assist in development of program. Benefits can be suspended for noncooperation with vocational rehabilitation.
New Jersey	No fund established.		PT disability benefits may be stopped after 450 weeks unless the worker has submitted to physical or educational retraining.
New Mexico	No fund established.		
New York	$2,000 no-dependency death cases.	Up to $30 per week for rehabilitation maintenance.	The statute provides for direction by the State Education Department of the rehabilitation process. It does not provide a role for the Department of Labor.
North Carolina	No fund established.		Insurer must furnish rehabilitation services required to lessen disability. Employee's unreasonable refusal of services ordered by Commission results in loss of compensation for period of refusal.
North Dakota	Benefit fund	Rehabilitation allowance plus supplies, tuition, fees and books in lieu of and equal to compensation for up to 2 years, plus 25% if maintaining 2 domiciles.	Bureau through its Director of Rehabilitation provides retraining. Employee's unreasonable refusal to cooperate shall forfeit compensation. If worker refuses rehabilitation, no benefits will be provided. If refusal continues for 60 days, eligibility for benefits ends.
Ohio	State Insurance Fund.	Same as for TT disability, minimum 50% of the SAWW, for 6 months (renewable).[10]	Rehabilitation Division is within the Bureau of Workers' Compensation and may make all necessary expenditures, medically including treatment of non-occupational conditions inhibiting return to work.
Oklahoma	No fund established.	Board, lodging, travel, tuition and books.	Court may order necessary rehabilitation up to 52 weeks; may also order additional 52 weeks.
Oregon	No fund established.	TT during training. Necessary expenses including tuition, books, some travel costs and tools.	Insurer must provide vocational assistance to workers who cannot return to work at a wage within 20% of wage at injury.
Pennsylvania	No fund established.[11]	State Board of Vocational Rehabilitation may provide cash payments for living expenses.	State Board of Vocational Rehabilitation may provide physical restoration, training, support and job placement services.
Puerto Rico	No fund established.	Administrator may grant $65 weekly up to 26 weeks.	Rehabilitation center provides physical, medical and rehabilitation services.
Rhode Island	Workers' Compensation Administrative Fund.	Board, lodging and/or travel.	The Department operates the Dr. John E. Donley Rehabilitation Center. Compensation suspended for willful refusal of suitable employment or rehabilitation.
South Carolina	No fund established.		No specific statutory provision.
South Dakota		Compensation paid up to 60 days, if pursuing rehabilitation.	TT during period of approved vocational rehabilitation.
Tennessee	No fund established.		Division of Workers' Compensation refers feasible cases to Department of Education pursuant to plan providing full or partial recovery of expenses from employer or insurer.
Texas	No fund established.		Insurer furnishes necessary medical care and services for physical rehabilitation. Commission may notify employees of vocational rehabilitation services through the Texas Rehabilitation Commission and private providers. Commission keeps registry of private providers of rehabilitation services.
Utah	Voluntary by insurance carrier.		If worker cannot be rehabilitated, worker receives benefits for life from employer or insurance carrier, minimum is 36% of the current SAWW.
Vermont	No fund established.	Board, lodging, travel, books and tools.	Commissioner may order vocational rehabilitation services. If employee refuses, compensation may be suspended.

Chart X — Rehabilitation Of Disabled Workers

Jurisdiction	Source Of Fund	Maintenance Allowance	Special Provisions
Virgin Islands	Government Insurance Fund.	Board, lodging and travel.	Department of Labor cooperates with Department of Human Services. Income benefits increased to 75% SAWW. Income benefits during rehabilitation suspended for employee's refusal to accept vocational rehabilitation. See Chart V — Total Disability Benefits.
Virginia	No fund established.		Commission may award compensation, medical care and vocational rehabilitation. Employer may be required to furnish and maintain wheelchairs, bedside lifts, adjustable beds and make alterations to home, maximum $25,000. Employee's unreasonable refusal may suspend compensation.
Washington	No fund established.	Compensation; board, lodging, travel, books, equipment and child care allowance, up to 52 weeks (maximum $3,000).	Supervisor may extend period for another 52 weeks. Department operates a Rehabilitation Center and pays maintenance and employer's cost of job modification.
West Virginia	State fund used, no special account.	Up to $10,000 (includes tuition, books, supplies, travel, lodging and tools). No limit on physical rehabilitation costs. TT disability payments if totally disabled.	Fund-employed Rehabilitation Counselors provide referrals and direct services. Direct job placement emphasized, but training considered on basis of need. Short term training preferred. Longer programs approved when no other employment alternatives available, normally limited to 2 years.
Wisconsin	No fund established.	TT disability, travel and necessary maintenance if away from home.	80-week period may be extended with Division's permission if necessary. Division refers feasible causes to Department of Vocational Rehabilitation. If Department of Rehabilitation cannot serve the employee, employee is eligible to select a private specialist to provide vocational rehabilitation services.
Wyoming	Workers' Compensation Fund.		Work with Department of Vocational Rehabilitation after eligibility determination with individualized rehabilitation plan.
F.E.C.A.	Employees' Compensation Fund.	Up to $200 per month.	If person fails to undergo rehabilitation, administrator may reduce benefit if rehabilitation would have increased earnings.
Longshore Act	Special Fund	Up to $25 per week.	Surplus in Fund in any 1 year may be carried over. Appropriations authorized.
Alberta	Accident Fund	Discretion of Board.	Board operates physical rehabilitation center. Board may make necessary expenditures to aid rehabilitation and may provide vocational rehabilitation to a dependent spouse.
British Columbia	Accident fund.	Discretion of Board.	Rehabilitation Clinic established. Board may make necessary expenditures to aid rehabilitation and may provide vocational rehabilitation to a dependent spouse.
Manitoba	Accident fund	Discretion of Board.	Board may make necessary expenditures to aid rehabilitation. Spouse of deceased worker receiving monthly payments entitled to rehabilitation.
New Brunswick	Accident Fund	Discretion of Commission.	Commission operates physical rehabilitation center. Commission may make necessary expenditures to aid rehabilitation.
Newfoundland	Injury Fund	Discretion of Commission.	Commission may make expenditures as are necessary or expedient to aid a worker's rehabilitation.
Northwest Territories	Accident Fund	Discretion of Board.	Board may make necessary expenditures to aid rehabilitation.
Nova Scotia	Accident Fund	Discretion of Board.	Board may make necessary expenditures to aid rehabilitation.
Ontario	Schedule 1—Accident Fund Schedule 2—employer's individually	Discretion of Board.	No limit on amount in any 1 case or in any 1 year for rehabilitation. Surviving spouses entitled to same level of vocational rehabilitation as injured workers.
Prince Edward Island	Accident Fund	Discretion of Board.	Board may make necessary expenditures to aid rehabilitation.
Québec	Accident Fund. Individual liability for employers held personally responsible for the payment of benefits.	Discretion of Board.	Board may make necessary expenditures to aid rehabilitation and may provide vocational rehabilitation to a dependent spouse.
Saskatchewan	Injury Fund	As required by law.	Board provides on-the-job training, employer assistance, and physical and occupational therapy. Vocational training available for spouse; educational allowance for dependent children of fatally injured workers.
Yukon Territory	Compensation Fund	Discretion of Board.	Board may make necessary expenditures to aid rehabilitation.
Canadian Merchant Seaman's Act	No fund established	No specific statutory provision.	No specific statutory provision.

[1] **Alaska** — Stipend of 60% of spendable income kicks in when medical and PPI benefits are exhausted.
[2] **Florida** — Refusal to accept reemployment services deemed necessary by the judge of compensation claims or the Division may result in a 50% reduction in weekly compensation benefits for each week of refusal.
[3] **Indiana** — State Rehabilitation Services Board administers vocational rehabilitation programs. Compensation suspended for refusal of suitable employment by partially disabled claimant.
[4] **Maine** — Board may levy assessment on each insurer based on paid losses for previous year when amount in Fund is less than $500,000. Fund also receives sum equal to 100 times the AWW when work-related injury causes employee death and there are no dependents.
[5] **Maine** — Office of Rehabilitation may implement and pay for out of the Rehabilitation Fund a plan previously rejected by the employer. If the plan proves successful, the Fund may assess the employer up to 180% of the plan implementation costs.
[6] **Massachusetts** — If insurer refuses payment, rehabilitation paid by Trust Fund. If rehabilitation is successful, insurer assessed no less than twice the cost of rehabilitation.
[7] **Minnesota** — Surviving spouse may request rehabilitation.
[8] **Minnesota** — Expenses of rehabilitation is borne by employer. Vocational rehabilitation is separated from physical rehabilitation. Physical rehabilitation is considered a medical expense.
[9] **Missouri**— Refusal of employee to undergo rehabilitation results in 50% reduction in TT or TP benefits.
[10] **Ohio** — If claimant returns to lesser-paying job while in rehabilitation, wage loss compensation of difference between wage at time of injury and wage at job while in rehabilitation program can be paid.
[11] **Pennsylvania** — Funded by federal and state sources.

Appendix E

Second Injury Funds

Reprinted with permission from the Chamber of Commerce of the United States from the 1995 Edition of *Analysis of Workers' Compensation Laws*. *Analysis of Workers' Compensation Laws* may be ordered by calling 1-800-638-6582.

Chart XIII — Second Injury Funds

Jurisdiction	Injuries Covered	Payable By Employer	Payable By Fund	Sources Of Fund	Special Provisions
Alabama[2]	Second Injury Fund eliminated.				
Alaska	Second injury which added to preexisting permanent physical impairment results in substantially greater disability than from second injury alone.	Disability caused by second injury up to 104 weeks.	Compensation in excess of 104 weeks.	Up to 6% of compensation payable to fund; percentage varies from 0% to 6% depending on fund balance. $10,000 in no-dependency death cases; civil penalties.	"Physical impairment" as listed or would support an award of 200 weeks or more.
American Samoa	Second injury which combined with prior permanent impairment results in death or compensable disability greater than from second injury alone.	Benefits for first 104 weeks.	Benefits beyond first 104 weeks.	$1,000 in no-dependency death cases, plus fines and penalties.	Employer must have prior knowledge of disability.
Arizona	Second injury which added to a preexisting work-related disability or a preexisting physical impairment not industrially related (25 types of handicaps as listed by statute) results in disability for work.	Disability caused by second injury.	Employer and special fund are equally liable for remaining difference between compensation payable for second injury and compensation for combined disability.	1.5% of all premiums and costs of self-insurance. Commission may allocate up to .5% of yearly premiums to special fund to keep fund actuarially sound.	Employer must have knowledge of non-industrial physical impairment. Payments are also made from the fund for vocational rehabilitation, claims against non-insured employers, insolvent carriers, supportive medical care.
Arkansas	Second injury which added to previous PP disability or impairment results in additional disability or impairment greater than from second injury alone.	Disability caused by second injury.	Difference between compensation payable for second injury and permanent disability.	$500 in no-dependency death case to be paid to the Permanent Total Disability and Death Fund. A portion of premium tax allocated to the Second Injury Fund and Death and Permanent Total Disability Trust Fund.	Employer liable for combined disability of both injuries in same employment.
California	Second permanent partial injury which added to preexisting PP disability results in 70% or more permanent disability. Second injury must account for 35%.[3]	Disability caused by second injury.	Difference between compensation payable for second injury and permanent disability.	Legislative appropriations and $50,000 in each no-dependency death case or unpaid balance.	Payments are made by State Compensation Insurance Fund.
Colorado[4]	Second injury which added to preexisting PP disability results in PT disability. Includes specified occupational diseases with multiple exposure— asbestosis, silicosis, anthracosis and disease from radiation exposure.	Disability caused by most recent injury. In addition, in occupational disease cases, fund pays all medical and all compensation above $10,000.	Difference between compensation payable for most recent injury and PT disability.	$15,000 in no-dependency or partial dependency cases, and fee on 1% of premiums received by insurance carriers and equivalent charge on self-insurers.	If employee obtains employment while receiving compensation from second-injury fund, fund compensates at rate of 1/2 of employee's AWW loss during employment.
Connecticut	Second injury or disease which added to preexisting injury, disease or congenital, causes results in permanent disability greater than from second injury alone.	Benefits for first 104 weeks, less compensation payable for prior disability.	Benefits beyond first 104 weeks, less compensation payable for prior disability.	Tax equal to 5% of compensation paid by carriers and self-insurers during preceding calendar year plus fines.	Tax imposed each time fund balance is reduced to $1,000,000.
Delaware	Second injury or disease which added to existing permanent injury from any cause results in PT disability.	Disability caused by second injury.	Difference between compensation payable for second injury and permanent disability.	Tax of 2% of premiums received by insurance carriers and equivalent charge on self-insurers.	Payments suspended when fund reaches $750,000 and resumed when below $250,000.
District of Columbia	Second injury or disease which added to preexisting injury, disease or congenital causes results in permanent disability greater than from second injury alone.	Disability caused by second injury for first 104 weeks and all medicals.	Difference between compensation payable for second injury and permanent disability.	$5,000 in no-dependency death cases or unpaid awards. Pro rata assessments upon carriers and self-insurers based on paid losses. Fines and penalties.	For injuries occurring on or after 3/16/91, the fund will only reimburse lost wages. The employer will remain liable for medical cost.
Florida	Second injury or disease which merges with previous permanent physical impairment and results in substantially greater disability than from the second injury alone if injury or occupational disease and the permanent physical impairment is one of the listed diagnoses in the statute.		Fund reimburses employer for 50% of all benefits paid over and above $10,000. Preferred worker program (previously injured worker with impairment) shall reimburse an employer for the cost of insurance premium related to the preferred worker for up to 3 years of continuous employment.	Pro rata annual assessment upon net premiums of insurers and self-insurers.	Assessment must equal average of the sum of disbursements for the previous 3 calendar years, plus 2 times the disbursement of the most recent calendar year. Employee must have knowledge of pre-existing conditions and that physical impairment is limited to those listed in statute.
Georgia	Second injury or disease which merges with prior permanent physical impairment and results in greater disability than from second injury alone.	Disability caused by second injury for first 104 weeks.	Employer reimbursed for 50% of medical and rehabilitation expenses in excess of $5,000 up to $10,000, and 100% of medical and rehabilitation expenses in excess of $10,000, plus income benefits beyond 104 weeks.	Assessments on carriers and self-insurers proportionate to 175% of disbursements from fund to annual compensation benefits paid, less net assets in fund. In no-dependency death cases, the lesser of 1/2 of benefits payable or $10,000.	Employer must have prior knowledge of impairment. Assessments may be reduced or suspended when no funds are needed.
Guam	Second injury which combined with a previous disability causes PT disability.	Disability caused by second injury.	Difference between compensation payable for second injury and compensable disability.	State fund (appropriation).	
Hawaii	Second injury which added to preexisting disabilities results in greater permanent disability, PT disability or death.	Disability benefits for first 104 weeks.	Compensation beyond first 104 weeks.	25% of 312 x effective maximum weekly benefit rate in no-dependency death cases; unpaid balance of compensation due in PT and PP disability cases if no dependents; special assessment on insurers and self-insurers.	

CHART XIII — SECOND INJURY FUNDS

JURISDICTION	INJURIES COVERED	PAYABLE BY EMPLOYER	PAYABLE BY FUND	SOURCES OF FUND	SPECIAL PROVISIONS
Idaho	Second injury which combined with prior permanent physical impairment results in PT disability.	Disability caused by second injury.	Difference between compensation payable for second injury and permanent disability.	Amount equal to 5% of all benefits except temporary disability income benefits and accrued medical benefits and $10,000 in no-dependency death cases.	When fund exceeds $2,000,000, excise shall be reduced to 4% and increased to 5% when below $2,000,000.
Illinois	Second injury involving loss or loss of use of major members or eye which added to preexisting loss of member results in PT disability.	Disability caused by second injury.[5]	Difference between compensation payable for second injury and PT disability.	Semi-yearly employer payment of .125% of compensation payments.	When fund reaches $500,000, amount payable into fund reduced by 1/2; payments cease when fund reaches $600,000. When fund reduced to $400,000, payment of 1/2 amount required. When fund reaches $300,000, payment of full amount resumed.
Indiana	Second injury involving loss or loss of use of hand, arm, foot, leg or eye which added to preexisting loss or loss of use of member results in PT disability.	Disability caused by second injury.	Difference between compensation payable for second injury and PT disability.	1% of compensation paid by insurers and self-insurers during preceding calendar year.	Payment suspended when fund reaches $400,000.
Iowa	Second injury involving loss or loss of use of member or eye which added to preexisting loss of use of member results in permanent disability.	Disability caused by second injury.	Difference between compensation payable for second injury and permanent disability, less value of previous loss of member or organ.	$4,000 in dependent death cases; $15,000 in no-dependent death cases; any contributions by the United States; payments due but not paid to non-resident alien dependents; and sums recovered from third parties.	Payments to the fund are suspended when fund reaches $1,000,000, resumed when below $5,000,000. A surcharge on weekly benefits paid by insurers and self-insurers is imposed to collect up to $870,000 for the Second Injury Fund.
Kansas	Second injury related to 17 types of handicap as listed in statute—any physical or mental impairment.	Difference between fund payment and maximum award.	Compensation to the extent pre-existing handicap contributed to second injury.	$18,500 from employer in no-death cases, and pro rata annual assessment upon carriers and self-insurers based on losses.	Compensation for preexisting handicaps contribution to second injury only applicable to claims arising from injuries prior to 7/1/94
Kentucky	Second injury or disease which added to prior disability or condition results in permanent disability greater than from second injury alone.	Disability caused by second injury.	Difference between compensation payable for second injury and greater disability, less amount paid for prior injury, but 50% or less of total income benefit awarded.	16.9% on all policies and an additional tax of 47% on premiums for policies issued to coal companies.	
Louisiana	Second injury which combined with prior PP disability results in disability greater than from second injury alone, or in death.[6]	Total disability benefits for first 104 weeks; in death cases, first 175 weeks; 50% of medical benefits which exceed $5,000 but are less than $10,000, and 100% thereafter is how much employer is reimbursed.	Employer is reimbursed for all weekly compensation payments payable after first 104 weeks, or 175 weeks concerning a death.	2 and 3/4% assessment on carriers and self-insurers, minimum $10.	Assessments reduced at discretion of the Board within 30 days written notice before assessment is due.
Maine	Second Injury Fund eliminated.				
Maryland	Second injury which combined with a pre-existing permanent impairment due to accident, disease or congenital condition results in a greater combined disability constituting a hindrance to employment.	Disability caused by second injury.	If permanent disability exceeds 50% of the body as a whole, employee is entitled to additional compensation for the full disability from the "Subsequent Injury Fund". Prior and second injury must each be compensable for at least 125 weeks.	6-1/2% to the Second Injury Fund and 1% to the Uninsured Employer's Fund of compensation on all awards and settlement agreements.	
Massachusetts	Second injury which added to preexisting physical impairment results in substantially greater disability or death.	Benefits for first 104 weeks.	Employer reimbursed for up to 75% of benefits after first 104 weeks.	Assessments on employers.	Pro rata assessment based on losses paid during preceding year by carriers and self-insureds.
Michigan	Second injury involving loss of member or eye, which added to preexisting loss of member results in PT disability.	Disability caused by second injury.	Difference between compensation for second injury and PT disability. Benefits for employee with more than 1 job but for whom injury occurred on job which represented less than 80% AWW. For workers certified "vocationally handicapped," fund pays benefits after 52 weeks.	Assessments on carriers and self-insurers proportionate to 175% of disbursements from fund to annual compensation benefits paid.	Fund is credited with any balance in excess of $200,000.[7]
Minnesota	Second injury that results in substantially greater disability than would have resulted from second injury alone.	Disability caused by second injury.	Employer reimbursed for disability after 52 weeks, medical after $2,000. If second injury results in PP disability, fund pays wage replacement benefits beyond deductible.	$25,000 in no-dependency death cases, 31% of compensation for current indemnity payments, which finances all fund obligations; assessment based on various factors for injuries occurring after 1/1/84; certain penalties.	Commissioner determines assessment base and rate dependent on fund's financial position and increasing up to 12% annually.[8]
Mississippi	Second injury involving loss or loss of use of member or eye, which added to preexisting loss or loss of use of member or eye results in PT disability.	Disability caused by second injury.	Difference between compensation payable for second injury and permanent disability.	$500 in no-dependency death cases; $300 in dependency cases. Commission may transfer up to $200,000 from Administrative Expense Fund.	Payments suspended when fund reaches $350,000 and resumed when fund reduced to $150,000.

CHART XIII — SECOND INJURY FUNDS

JURISDICTION	INJURIES COVERED	PAYABLE BY EMPLOYER	PAYABLE BY FUND	SOURCES OF FUND	SPECIAL PROVISIONS
Missouri	Second injury, which combined with permanent industrial disability, results in a greater combined PP or PT disability.	Disability caused by second injury.	Difference between compensation payable for second injury and compounded disability.	Surcharge on all workers' compensation premiums not to exceed 3% of premiums, paid by all insureds and self-insurers—according to statutory formula.	Surcharge suspended when balance of fund exceeds 110% of the monies projected paid from Second Injury Fund in the ensuing calendar year and resumed when balance of fund is less than 110% of monies projected to be paid from the Second Injury Fund.
Montana	Any new compensable injury following certification.	Insurer liable for payment of benefits for first 104 weeks.	Employer reimbursed after first 104 weeks.	$1,000 paid by employers, insurers or accident fund in every death case. Carriers and self-insurers assessed up to 5% of losses paid in preceding year.	Department must certify worker as vocationally handicapped.
Nebraska	Second injury which combined with preexisting disability causes substantially greater disability. Preexisting disability must support 25% earnings loss or 90 weeks of benefits.[1]	Disability caused by second injury.	Difference between compensation payable for second injury and the total resulting disability.	2% assessment of benefits paid by carriers or self-insurers in the state in the prior year.	Payments suspended when fund reaches $800,00. Assessment (2%) when fund reduced to $400,000. Employer must have knowledge of preexisting permanent disability.
Nevada	Second injury which combined with any previous permanent physical disability causes substantially greater disability.[9]		Compensation allocated between self-insured employer and fund. State Fund does not contribute, nor does it receive funds.	Self -insured employers contribute to and may receive payments from Subsequent Injury Fund in state treasury.[10]	Compensation claim considered "excess loss" in calculation of employer's experience rating. Employer must prove knowledge of prior impairment.
New Hampshire	Second injury which combined with any preexisting disability results in greater disability.[1]	Benefits for first 104 weeks.	Employer reimbursed after first 104 weeks and for 50% of anything over $10,000 during first 104 weeks.	Assessment against carriers and self-insurers proportional to total benefits paid by all carriers.	Employer who undertakes job modifications to retain injured worker is reimbursed 50% of cost of modification from the fund, not to exceed $5,000 per employee per year.
New Jersey	Second injury which in combination with preexisting partial disability, compensable or not, totally disabled employee.	Disability caused by compensable injury.	Difference between compensation payable for second injury and preexisting disability.	Annual surcharge on policyholders and self-insured employers of pro rata percentage of 150% of payments estimated to be paid from fund during forthcoming year. Annual surcharge paid quarterly.	When fund balance exceeds, $1,250,000, up to $50,000 per year may be applied toward administration costs of Division.
New Mexico	Second injury which added to preexisting disability greater than from second injury alone; or second injury resulting in death.	Liability apportioned by Workers' Compensation Administration determination.	Liability apportioned by Workers' Compensation Administration determination.	$1,000 in no-dependency death cases. Employer or insurer pays quarterly assessment up to 3% of compensation paid during quarter, exclusive of attorney's fees.	Employers and insurers have 2 year statute of limitations, from date of notice or knowledge of claim, for claims against fund.
New York	Second injury where employee has a preexisting permanent physical impairment resulting in a permanent disability caused by both conditions that is materially and substantially greater than that which would have resulted from the second injury alone.	Benefits for first 104 weeks.	Employer reimbursed after first 104 weeks. Fund also pays any additional benefits due to an employee who was working in concurrent employments when injured.	Assessment against carriers and self-insurers proportional to compensation payments made by all carriers.	Employer or insurer pays award and medical expenses, but is reimbursed from special disability fund for benefits after first 104 weeks.
North Carolina	Second injury involving loss of member or eye which added to preexisting injury results in PT disability, provided the original and increased disability were each 20% of the entire member.[11]	Disability caused by second injury.	Difference between compensation payable for second injury and PT disability.	Assessments against employer or insurer for each PP disability, up to $250 for a minor member or $750 for 50% or more loss or loss of use of each major member — back, foot, leg, hand, arm, eye, or hearing.	
North Dakota	Second injury or aggravation of any previous injury or condition which results in further disability.	Disability caused by second injury.	Percent attributable to aggravation or second injury.	Benefit fund.	Compensation in excess of amount chargeable to second injury is charged to general fund. Prior registration into the fund is not required or accepted.
Ohio	Second injury which aggravates preexisting disease or condition (25 types of handicaps as listed by statute), resulting in death, temporary or PT disability, and disability compensable under a special schedule.[12]	Disability attributable to injury or occupational disease sustained in employment.	Amount of disability or proportion of cost of death award determined by Bureau to be attributable to employee's preexisting disability. The Bureau's decision is appealable to the Industrial Commission.	Reserve set aside out of statutory surplus funds.	In the case of a self-insuring employer, excess payments made from surplus fund, though self-insuring employers may opt not to participate in the handicap reimbursement fund. By rule of Commission in the case of State Fund employer, compensation excess of amount chargeable to second injury is charged to surplus fund.
Oklahoma	Second injury to "physically impaired person",[13] which causes injury to the body as a whole, or to a major member combined with disability to body as a whole. Must exceed 40% to body as a whole.	Disability caused by latest injury.	Difference between compensation payable for prior injuries and compensation for combined injuries.	5% of permanent disability losses by carrier, state fund, and self-insurers and 5% of awards for permanent disability by injured worker.	PT awards are payable by the fund for five years or until 65, whichever is longer.

Chart XIII — Second Injury Funds

Jurisdiction	Injuries Covered	Payable By Employer	Payable By Fund	Sources Of Fund	Special Provisions
Oregon	Any new compensable injury sustained by an injured worker within 3 years from the hire date as a Preferred Worker through the Reemployment Assistance Reserve.	None	Employers hiring a Preferred Worker are exempt from paying premiums and premium assessments on the worker for 3 years from hire date. Reserve reimburses all claim costs incurred by the worker for any new compensable injury within 3-year period. Other return-to-work incentives include work-site modification up to $25,000, wage subsidy of 50% up to 6 months and necessary purchase for obtained employment.	Worker and employer each pay 1.7 cents per hour into Reserve. Reserve also funds three other programs.	Reimbursement from Reserve subject to funds available. Decisions regarding eligibility and extent of assistance not reviewable. Settlement of claim requires department approval if reimbursement involved.
Pennsylvania	Second injury involving loss or loss of use which added to pre-existing loss or loss of use of member results in PT disability.[14]	Scheduled benefits as a result of second injury.	Remaining compensation due for total disability.	Assessment against carriers and self-insurers proportional to compensation payments.	Payments are made directly by the Department.
Puerto Rico	Second injury which aggravates or augments any former disability.		Job injury not caused by work accident is compensated in addition to second injury. Compensation for prior job injury is deducted from compensation payable for total disability, except where combined injury results in PT disability, which is compensated as such.	Insurance premiums defray work-related injuries; however, previous impairments are covered from the Fund for Catastrophes which does not affect employers accident experience ratings or premiums.	The difference between expenditures by the Industrial Commission and the Manager of the State Insurance Fund and their maximum budget allotment are placed in the Reserve Fund for catastrophes except for medical expense surpluses; maximum $1 million.
Rhode Island	Second injury which merges with preexisting work-related disability resulting in greater disability or death.	Benefits for first 26 weeks.[15]	Employer reimbursed for compensation and medical after first 26 weeks.[15]	Assessment payable by insurers (and group self-insurers) is 4.75%[16] of net premiums written for workers' compensation; self-insurers pay 4.75[16]of premium they would have paid to be insured; $1,500 in no-dependency death cases; also certain penalties.	Employer must prove knowledge of prior injury unless employee failed to disclose. Insurers who discontinue writing compensation policies are obligated to pay reduced assessments for 6 years.
South Carolina	Second injury which added to any previous permanent physical impairment results in substantially greater disability or death.	Disability caused by second injury for first 78 weeks' compensation and medical care.	Employer reimbursed for all benefits after 78 weeks, plus 50% of medical payments over $3,000 during first 78 weeks.	Pro rata assessments on carriers and self-insurers based on losses paid. In no-dependency deaths, unpaid benefits to fund.	Employer must prove prior knowledge of impairment or that worker was unaware of impairment.[17] Any claim against the Fund must be filed with the Fund prior to payment of 78 weeks of benefits.
South Dakota	Second injury, which combined with any preexisting disability, results in additional PP or PT disability or death.	Disability caused by second injury.	Difference between compensation payable for second injury and compensation for combined injuries.	Carriers and self-insurers assessed 4% of losses paid during preceding year and $500 in no-dependency death cases.	Payments suspended at $200,000, resumed at $100,000. Any claim against the Fund must be filed with the Division of Insurance within 90 days of the subsequent injury.
Tennessee	Second injury involving loss or loss of use of member or eye, which added to preexisting loss or loss of use of member results in PT disability.[18]	Disability caused by second injury.	Benefits in excess of 100% total disability to body as a whole.	50% of revenues from the 4% premium tax on insurers and self-insurers.	
Texas	Subsequent compensable injury combined with the effects of a previous injury entitles employee to lifetime income benefits.	Benefits which would accrue if only the subsequent and not previous injury occurred.	Balance of lifetime income benefits due.	Maximum $150,424[19] payable into fund in each no-dependency death case.	Workers' Compensation Commission had right of subrogation to recover claims and attorney's fees paid from Second Injury Funds.
Utah	Second injury which combined with a previous permanent incapacity due to accident, disease or congenital condition results in PT disability.	Employer pays first 6 years of PT unless a 10% preexisting condition, then employer pays first $20,000 of medical benefits and first 3 years of PT.	50% of medical expenses in excess of $20,000 and PT disability compensation after initial 3 year period. Balance of lifetime benefits after initial 3 or 6 years.	Up to 7.25% tax on insurers and self-insurers.	If employee is permanently and totally disabled, employer or insurance carrier credited for all prior payments of TT, TP, and PP disability compensation.
Vermont	Second injury involving loss of use of member or eye which added to previous disability results in PT disability.	Disability caused by second injury.	Difference between compensation payable for second injury and PT disability.	$500 in no-dependency death cases.	Payments suspended until sufficient funds become available.
Virgin Islands	Second injury which combined with prior impairment results in death or compensable disability greater than from second injury alone.	None. Employer's experience rating affected by disability after 104 weeks.	All benefits.	Premiums paid by employers by classification and experience, plus fines, penalties and interest to Governmental Insurance Fund.	Employer must have prior knowledge of disability.
Virginia	Second injury involving 20% loss or loss of use of member or eye which added to preexisting disability of 20% or more results in total or partial disability.	Disability caused by second injury.	Employer reimbursed for compensation after all other compensation has expired plus up to $7,500 each for medical and vocational rehabilitation expenses.	1/4% premium tax on carriers and self-insurers.	Payments suspended at $250,000 and resumed at $125,000.
Washington	Second injury or disease which added to preexisting injury or disease results in PT disability or death.	Disability caused by second injury.	Difference between charge assessed against employer at time of second injury and total pension reserve.	Transfer of not more than cost from accident fund to second injury account. Self-insurers pay proportional to claims against self-insurers.	Preferred workers[20] have all benefits for claims arising within 3 years of new employment paid Second Injury Fund. The Second Injury Fund also covers job modification costs resulting from on-the-job injuries.

CHART XIII — SECOND INJURY FUNDS

Jurisdiction	Injuries Covered	Payable By Employer	Payable By Fund	Sources Of Fund	Special Provisions
West Virginia	Second injury which combined with a definitely ascertainable physical impairment caused by prior injury results in PT disability.	Disability caused by second injury.	Remainder of the compensation that would be due for PT disability.	Self-insureds in mining industry pay 30% of the manual rate; high risk industry pay 20%, all other self-insureds pay 17%.	Self-insured employer who has elected not to pay into the fund liable for full compensation of PT disability from combined effect of a previous injury and a second injury.
Wisconsin	Second injury with permanent disability for 200 weeks or more with a preexisting disability of an equal degree or greater.	Disability caused by second injury.	Disability caused by lesser of 2 injuries. If the combined disabilities result in PT disability, fund pays the difference between compensation payable for second injury and PT disability.	$5,000 in death cases; $7,000 for loss of hand, arm, foot, leg or eye. 100% of death benefit in no-dependency death cases.	
Wyoming			Not applicable.		
Longshore Act	Second injury resulting in PP disability which added to preexisting injury results in PT disability or greater PP disability.	Disability caused by second injury for first 104 weeks.	Balance of compensation after 104 weeks.	$5,000 in no-dependency death cases or unpaid awards. Pro rata assessments based on losses paid. Fines and penalties.	
Alberta	All enhanced disabilities due to the aggravation of preexisting condition.	No	Yes	Accident fund	
British Columbia	All enhanced disabilities by reason of a pre-existing disease, condition or disability.	No	Yes	Accident fund	
Manitoba	Cost relief fund includes preexisting or underlying conditions, occupational diseases with exposure outside Manitoba, loss of earning from an employment other than worker's employer and some increase in benefits due to recurrences or age or apprenticeship of worker, or other cirumstances which the WCB determines would unfairly burden a particular class, sub-class, group, or sub-group.	No	Yes	Accident fund	
New Brunswick	Second injury coupled with other prior injuries or disabilities.	No	Yes	Reserve fund	
Newfoundland	All enhanced disabilities because of similar or other disabilities.	No	Difference between compensation payable for second injury and final result of disablement.	Reserve fund	
Northwest Territories	All disabilities due to preexisting disease, condition or disability.	No	Difference between second injury and total cost.	Accident fund	
Nova Scotia	Injury that aggravates, activates or accelerates a disease or disability existing prior to the injury; or injury that results in injury or disease caused partly by employment and partly by other causes.	No	Disability attributable to second injury.	Accident fund	Board has authority to establish second injury fund.
Ontario	All disabilities caused or enhanced due to preexisting diseases, condition or disability.	No[21]	Determined by WCB: apportionment usually 50% but may range from 25% -100% of cost, subject to approval.	Accident Fund	Not restricted to permanent disability cases.
Prince Edward Island			No specific statutory provision.		
Québec				Commission has authority to establish Second Injury Fund.	
Saskatchewan	All enhanced disabilities due to preexisting disease, condition or disability.	No	Difference between second injury and total cost.	Injury fund	
Yukon Territory	All enhanced disabilities because of similar or other disabilities.	No	Yes	Compensation Reserve Fund for enhanced disabilities. Assessment on employers' annual payroll.	
Canadian Merchant Seaman's Act			No statutory provision.		

[1] In death cases it must be established that either the injury or the death would not have occurred except for such preexisting permanent physical impairment. "Permanent physical impairment" means any permanent condition due to previous accident, disease, or congenital condition which is likely to be a hindrance to employment.

[2] **Alabama** — Second Injury Trust Fund repealed on 5/19/92. An amount is included in the annual workers' compensation budget which shall be allocated for the specific and exclusive purpose of paying only benefits to the claimants who have qualified to receive benefits from the Second Injury Trust Fund on 5/19/92.

[3] **California** — Second injury must account for 35% unless prior disability involved a major member and second injury was to opposite and corresponding member and accounts for at least 5%. No benefits payable for subsequent unrelated noncompensable injury.

[4] **Colorado** — Fund is closed to injuries occurring on or after 7/1/93 and for occupational diseases occurring on or after 4/1/94.

[5] **Illinois** — Employer is liable in full if second injury is permanent and total without relation to prior injury.

[6] **Louisiana** — PP disability means any permanent condition due to injury, disease or congenital causes which is likely to be a hindrance to employment. Certain scheduled conditions presumed to be PP disabilities if employer had prior knowledge.

[7] **Michigan** — Compensation to certified vocationally handicapped persons payable from fund after 52 weeks.

[8] **Minnesota** — If injury, disability or death would not have occurred but for the preexisting impairment, the fund pays all benefits (except for a cardiac condition, impairment of at least 10% of the whole man, or as prescribed by the rule). Second-Injury Fund has been abolished after 6/30/92. Dates of injury prior to 7/1/92 are reimbursable and assessable.

[9] **Nevada** — Preexisting disability must support a rating of 6% or more of the whole person based on A.M.A. guides, which is likely to be a hindrance to employment.

[10] **Nevada** — Fund is composed of assessments, penalties, bonds, securities and all other property collected by administrator section of Industrial Insurance Regulation.

[11] **North Carolina** — Epilepsy is considered a prior permanent disability.

[12] **Ohio** — Does not apply to compensation for TP or percentage of PP disability.

[13] **Oklahoma** — Physically impaired person — by accident, disease, birth, military action, or any other cause, has suffered the loss of sight of 1 eye, the loss by amputation of the whole or a part of, or loss of use of, a member such as is obvious from observation or examination by an ordinary layman.

[14] **Pennsylvania** — Benefits under the Subsequent Injury provision are only payable with respect to subsequent loss or loss of use of 1 hand , 1 arm, 1 foot, 1 leg or 1 eye.

[15] **Rhode Island** — For claims filed after 9/1/90. Employer reimbursed after first 52 weeks for injuries between 5/18/85 and 8/31/90; and before 5/18/85, after 104 weeks.

[16] **Rhode Island** — The percentage of assessment is determined by the Director on or before each March 15.

[17] **South Carolina** — Permanent physical impairment means any permanent condition due to injury, disease or congenital causes which is likely to be a hindrance to employment. Certain scheduled conditions are presumed to be permanent physical impairments if employer had prior knowledge.

[18] **Tennessee** — Also covers death and disablement resulting from injuries of an epileptic seizure occurring on or after 7/1/85.

[19] **Texas** — 360 times maximum weekly benefit.

[20] **Washington** — Preferred workers defined as workers who must change jobs due to effect of an industrial injury or illness.

[21] **Ontario** — Only applies to Schedule 1 employers and Schedule 2 employers are not eligible for Second-Injury Fund.

Appendix F

Notice to Employers—Claim Findings

Notice to Employers

The following states and organizations require 30 days notification to the employer: Alaska, American Samoa, California, District of Columbia, Florida, Georgia, Guam, Louisiana, Minnesota, Mississippi, Missouri, Montana, New York, North Carolina, Oregon, Rhode Island, South Dakota, Tennessee, Texas, Virgin Islands, Virginia, and Wisconsin.

The Federal Employees' Compensation Act and the Longshoreman's Act also require 30 days notification.

The following states and groups have requirements of immediate to 48 hour notification: Arizona, Arkansas, Connecticut, District of Columbia Government Workers, Hawaii, Indiana, Kentucky, Massachusetts, Nebraska, Nevada, New Hampshire, South Carolina, Vermont, Washington, West Virginia and Wyoming.

The following states have notification time frames between 4 days and 30 days: Alabama, Colorado, Kansas, Maryland, New Jersey, New Mexico, and Pennsylvania.

The following do *not* require notification to employer: North Dakota, Ohio (except to self-insurers), and Puerto Rico.

The following require more than 30 days notification: Delaware, Idaho, Illinois, Iowa, Maine, Michigan, Oklahoma, and Utah.

Claim Filings

Most states require that the claim be filed within 1-2 years after the injury occurred or after the last payment to the employee. The following states have longer than 2 year requirements: Illinois (3 years), Massachusetts (4 years), Minnesota (3 years), Pennsylvania (3 years), and Rhode Island (3 years).

The Federal Employees' Compensation Act also requires (3 years).

The following have shorter time requirements: District of Columbia Government Workers (60 days), Kansas (200 days), Maryland (60 days), Nevada (90 days), Puerto Rico (5 days), Vermont (6 months), and Virgin Islands (60 days).

Appendix G

Employer's Report of Accidents—Injuries Covered

Employer's Report of Accidents

All but the following states require the employer to keep accident records: Arizona, Kansas, Maryland, Missouri, North Dakota, Rhode Island, Tennessee, Washington, and West Virginia.

Injuries Covered

The following states cover all injuries from the workplace: Arizona, Colorado (if more than 3 days lost time), District of Columbia, Florida, Georgia (if more than 7 days absence), Hawaii, Idaho (if more than 1 days absence), Maine (if requires physician service or 1 days lost work), Montana, Nebraska, Nevada, New Hampshire (if lost time or medical expense), New Jersey, Oklahoma (if lost time or treatment away from worksite), Oregon, Puerto Rico, South Carolina (if medical care needed), South Dakota (if treatment other than first aid or 7 days absence), Tennessee (if medical treatment needed), Utah (medical treatment needed), Virginia, Washington (medical care needed), and West Virginia.

The Longshoreman's Act also covers all workplace injuries.

The following states are as indicated:

Alabama — Death or disability exceeding 3 days
Alaska — Death, injury, disease, or infection
American Samoa — Injury or death
Arkansas — Injury or death
California — Death cases or serious injuries, 1 day or more of first aid, occupational diseases, or pesticide poisoning
Connecticut — Disability of 1 day or more
Delaware — Death cases or injuries requiring hospitalization, other injuries
Guam — Injury or death
Illinois — Death cases or serious injuries, disability over 3 days, permanent disability
Indiana — Disability of more than 1 day
Iowa — Disability of over 3 days
Kansas — Death cases, disability of over 1 day or more
Kentucky — Disability of more than 1 day
Louisiana — Lost time over 1 week, or death
Maryland — Disability over 3 days
Massachusetts — Disability of over 5 days
Michigan — Death cases, disabilities of 7 days or more, and specific losses
Minnesota — Death or serious injury, disability of 3 days or more

Mississippi — Disability of one day or working shift
Missouri — Death or injury
New Mexico — Compensable injuries
New York — Disability of 1 day or more requiring medical care beyond two first aid treatments
North Carolina — Disability of more than 1 day
North Dakota — No statutory provision
Ohio — Injuries causing 7 days or more total disability
Pennsylvania — Death cases, disability of 1 day or more
Rhode Island — Death cases, disability of 3 days or more, any claim resulting in medical care
Texas — Disability of more than 1 day or occupational disease
Vermont — Disability of 1 day or more or requiring medical care
Virgin Islands — Injury or disease
Wisconsin — Disability beyond 3-day waiting period
Wyoming — Compensable injuries

In addition, the Federal Employees' Compensation Act covers death or probable disability.

Appendix H

Directory of Workers' Compensation Offices

Directory of Workers' Compensation Offices

UNITED STATES

ALABAMA
Workmen's Compensation Division
Department of Industrial Relations
Industrial Relations Building
Montgomery, Alabama 36131
(205) 242-2868

ALASKA
Division of Workers' Compensation
Department of Labor
P.O. Box 25512
Juneau, Alaska 99802-5512
(907) 465-2790

Workers' Compensation Board
(same as above)

AMERICAN SAMOA
Workers' Compensation Commission
Office of the Governor
American Samoa Government
Pago, Pago, American Samoa 96799

ARIZONA
Industrial Commission
800 West Wahington
P.O. Box 19070
Phoenix, Arizona 85005-9070
(602) 542-4411

ARKANSAS
Workers' Compensation Commission
Justice Building
625 Marshall Street
Little Rock, Arkansas 72201
(501) 682-3930

CALIFORNIA
Department of Industrial Relations
Division of Workers' Compensation
455 Golden Gate Avenue
Room 5182
San Francisco, California 94102
(415) 703-3731

Workers' Compensation Appeals Board
455 Golden Gate Avenue
Room 2181
San Francisco, California 94102
(415) 703-1700

COLORADO
Division of Workers' Compensation
1120 Lincoln Street, 12th Floor
Denver, Colorado 80203
(303) 764-4321

Industrial Claims Appeals Office
1120 Lincoln Street, 7th Floor
Denver, Colorado 80203
(303) 894-2378

CONNECTICUT
Workers' Compensation Commission
1890 Dixwell Avenue
Hamden, Connecticut 06514

DELAWARE
Industrial Accident Board
State Office Building, 6th Floor
820 North French Street
Wilmington, Delaware 19801
(302) 577-2885

DISTRICT OF COLUMBIA
Department of Employment Services
Office of Workers' Compensation
1200 Upshur Street, NW
Washington, D.C. 20011
(202) 576-6265

FLORIDA
Division of Workers' Compensation
Department of Labor and Employment Security
301 Forrest Building
2728 Centerview Drive
Tallahassee, Florida 32399-0680
(904) 488-2548

GEORGIA
Board of Workers' Compensation
South Tower, Suite 1000
One CNN Center
Atlanta, Georgia 30303-2788

GUAM
Workers' Compensation Commission
Department of Labor
Government of Guam
P.O. Box 9970
Tamuning, Guam 96931-2970
(671) 646-9324

HAWAII
Disability Compensation Division
Department of Labor and Industrial Relations
P.O. Box 3769
Honolulu, Hawaii 96812
(808) 548-4131

Labor and Industrial Relations Appeals Board
888 Mililani Street
Room 400
Honolulu, Hawaii 96813
(808) 548-6465

IDAHO
Industrial Commission
317 Main Street
Boise, Idaho 83720
(208) 334-6000

ILLINOIS
Industrial Commission
100 West Randolph Street
Suite 8-200
Chicago, Illinois 60601
(312) 814-6555

INDIANA
Workers' Compensation Board
402 West Washington Street
Room W196
Indianapolis, Indiana 46204
(317) 232-3808

IOWA
Division of Industrial Services
Department of Employment Services
1000 E. Grand Avenue
Des Moines, Iowa 50319
(515) 281-5934

KANSAS
Division of Workers' Compensation
Department of Human Resources
800 SW Jackson St, Ste. 600
Topeka, Kansas 66612-1227
(913) 296-4000

KENTUCKY
Department of Workers' Claims
Perimeter Park West
1270 Louisville Road, Building C
Frankfort, Kentucky 40601
(502) 564-5550

LOUISIANA
Department of Labor
Office of Workers' Compensation Administration
P.O. Box 94040
Baton Rouge, Louisiana 70804-9040
(504) 342-7555

MAINE
Workers' Compensation Commission
Deering Building
State House Station 27
Augusta, Maine 04333
(207) 289-3751

MARYLAND
Workers' Compensation Commission
6 North Liberty Street
Baltimore, Maryland 21201
(410) 333-4700

MASSACHUSETTS
Department of Industrial Accidents
600 Washington Street, 7th Floor
Boston, Massachusetts 02111
(617) 727-4900

MICHIGAN
Bureau of Workers' Disability Compensation
Department of Labor
201 North Washington Square
P.O. Box 30016
Lansing, Michigan 48909
(517) 322-1296

Board of Magistrates
201 North Washington Square
P.O. Box 30016
Lansing, Michigan 48909
(517) 335-0642

Workers' Compensation Appellate Commission
Department of Labor
7150 Harris Drive
P.O. Box 30015
Lansing, Michigan 48909
(517) 335-5828

MINNESOTA
Workers' Compensation Division
Department of Labor and Industry
443 Lafayette Road
St. Paul, Minnesota 55155
(612) 296-6107

Workers' Compensation Court of Appeals
775 Landmark Towers
345 St. Peter Street
St. Paul, Minnesota 55102
(612) 296-6526

MISSISSIPPI
Workers' Compensation Commission
1428 Lakeland Drive
P.O. Box 5300
Jackson, Mississippi 39296-5300
(601) 987-4200

MISSOURI
Division of Workers' Compensation
Department of Labor and Industrial Relations
3315 West Truman Blvd
P.O. Box 58
Jefferson City, Missouri 65102
(314) 751-4231

Missouri Labor and Relations Commission
3315 West Truman Blvd.
P.O. Box 599
Jefferson City, Missouri 65102
(314) 751-2461

MONTANA
State Fund Insurance Company
P.O. Box 4759
Helena, Montana 59604-4759
(406) 444-6518

Workers' Compensation Court
P.O. Box 537
Helena, Montana 59624
(406) 444-7794

NEBRASKA
Workers' Compensation Court
State House, 12th Floor
P.O. Box 98908
Lincoln, Nebraska 68509-8908
(402) 471-2568

NEVADA
State Industrial Insurance System
515 East Musser Street
Carson City, Nevada 89714
(702) 687-5284

NEW HAMPSHIRE
Department of Labor
Divison of Workers' Compensation
State Office Park South
95 Pleasant Court
Concord, New Hampshire 03301
(603) 271-3171

NEW JERSEY
Department of Labor
Division of Workers' Compensation
Call Number 381
Trenton, New Jersey 08625-0381
(609) 292-2414

NEW MEXICO
Workers' Compensation Administration
1820 Randolph Rd, SE
P.O. Box 27198
Albuquerque, New Mexico 87125-7198
(505) 841-6000

NEW YORK
Workers' Compensation Board
180 Livingston Street
Brooklyn, New York 11248
(718) 802-6666

NORTH CAROLINA
Industrial Commission
Dobbs Building
430 North Salisbury Street
Raleigh, North Carolina 27611
(919) 733-4820

NORTH DAKOTA
Workers' Compensation Bureau
Russel Building, Hwy 83 North
4007 N. State Street
Bismarck, North Dakota 58501-0600
(701) 224-3800

Workers' Compensation Fund
(Same address and phone number as above)

OHIO
Workers' Compensation Board
30 West Spring Street
Columbus, Ohio 43266-0581
(614) 466-2950

Industrial Commission
(Same address as above)
(614) 466-3010

State Insurance Fund
(Same address and phone number as WCB)

OKLAHOMA
Oklahoma Workers' Compensation Court
1915 N. Stiles
Oklahoma City, Oklahoma 73105
(405) 557-7600

OREGON
Department of Insurance and Finance
21 Labor and Industries Building
Salem, Oregon 97310
(503) 378-4100

Workers' Compensation Board
480 Church Street SE
Salem, Oregon 97310
(503) 378-3308

PENNSYLVANIA
Bureau of Workers' Compensation
Department of Labor and Industry
1171 South Cameron Street, Room 103
Harrisburg, Pennsylvania 17104-2501
(717) 783-5421

Workers' Compensation Appeal Board
1171 South Cameron Street, Room 305
Harrisburg, Pennsylvania 17104-2511
(717) 783-7838

PUERTO RICO
Industrial Commissioner's Office
G.P.O. Box 4466
San Juan, Puerto Rico 00936
(809) 783-2028

RHODE ISLAND
Department of Workers' Compensation
610 Manton Avenue
P.O. Box 3500
Providence, Rhode Island 02909
(401) 272-0700

Workers' Compensation Court
1 Dorrance Plaza
Providence, Rhode Island 02903
(401) 277-3097

SOUTH CAROLINA
Workers' Compensation Commission
1612 Marion Street
P.O. Box 1715
Columbia, South Carolina 29202
(803) 737-5700

SOUTH DAKOTA
Division of Labor and Management
Department of Labor
Kneip Building, Third Floor
700 Governors Drive
Pierre, South Dakota 57501-2277
(605) 773-3681

TENNESSEE
Workers' Compensation Division
Department of Labor
501 Union Building
Second Floor
Nashville, Tennessee 37243-0661
(615) 741-2395

TEXAS
Workers' Compensation Commission
Southfield Building
4000 South IH 35
Austin, Texas 78704
(512) 448-77900

UTAH
Industrial Commission
160 East 300 South
Salt Lake City, Utah 84111
(801) 530-6800

VERMONT
Department of Labor and Industry
National Life Building
Drawer 20
Montpelier, Vermont 05620-3401
(802) 828-2286

VIRGIN ISLANDS
Department of Labor
Workers' Compensation Division
2131 Hospital Street
Christiansted, St. Croix, Virgin Islands
00820-4666
(809) 773-0471

VIRGINIA
Workers' Compensation Commission
1000 DMV Drive
P.O. Box 1794
Richmond, Virginia 23214
(804) 367-8600

WASHINGTON
Department of Labor and Industries
Headquarters Building
7273 Linderson Way, SW, 5th Floor
Olympia, Washington 98504
Director: (206) 956-4200
Deputy Director for Industrial Insurance: (206) 956-4209
Deputy Director for Policy and Planning: (206) 956-4205

Board of Industrial Insurance Appeals
410 West 5th Street
Capitol Center Building
Mail Stop FN 21
Olympia, Washington 98504
(206) 753-6823

WEST VIRGINIA
Bureau of Employment Programs
Workers' Compensation Division
Executive Offices
601 Morris Street
Charleston, West Virginia 25332-1416
(304) 558-0475

Workers' Compensation Appeals Board
601 Morris Street, Room 303
Charleston, West Virginia 25301

WISCONSIN
Workers' Compensation Division
Department of Industry, Labor, and Human Relations
201 East Washington Avenue
Room 161
P.O. Box 7901
Madison, Wisconsin 53707
(608) 266-9850

WYOMING
Workers' Compensation Division
Department of Employment
122 West 25th Street, 2nd Floor
East Wing, Herschler Building
Cheyenne, Wyoming 82002-0700
(307) 777-7159

Industrial Accident Fund
(Same address and phone number as above)

UNITED STATES (Misc.)

Department of Labor
Employment Standards Administration
Washington, D.C. 20210
(202) 219-6091

Office of Workers' Compensation Programs
(202) 219-7503

Division of Coal Mine Workers' Compensation
(202) 219-6692

Division of Federal Employees' Compensation
(202) 219-7552

Division of Longshore and Harbor Workers' Compensation
(202) 219-8572

Division of Planning Policy and Standards
(202) 219-7293

Benefits Review Board
800 K Street NW, Suite 500
Washington, D.C. 20001-8001
(202) 663-7500

Employees' Compensation Appeals Board
300 Reporters Building
7th & D Streets, S.W., Room 300
Washington, D.C. 20210
(202) 401-8600

CANADA

ALBERTA
Workers' Compensation Board
P.O. Box 2415
9912 107 Street
Edmonton, Alberta T5J 2D5
(403) 498-4000

BRITISH COLUMBIA
Workers' Compensation Board
P.O. Box 5350
Vancouver, British Columbia V6B 5L5
(604) 273-2266

MANITOBA
Workers' Compensation Board
333 Maryland Street
Winnipeg, Manitoba R3G 1M2
(204) 786-5471

Workers' Compensation Appeals Board
175 Hargrove
Room 311
Winnipeg, Manitoba R3C 3R8

NEW BRUNSWICK
Workers' Compensation Board
1 Portland Street
P.O. Box 160
Saint John, New Brunswick E2L 3X9
(506) 632-2200

NEWFOUNDLAND
Workers' Compensation Commission
P.O. Box 9000
Station B
St.John's, Newfoundland A1A 3B8
(709) 778-1000

NORTHWEST TERRITORIES
Workers' Compensation Board
P.O. Box 8888
Yellowknife, Northwest Territories X1A 2R3
(403) 920-3888

NOVA SCOTIA
Workers' Compensation Board
5668 South Street
P.O. Box 1150
Halifax, Nova Scotia B3J 2Y2
(902) 424-8440

Workers' Compensation Appeals Board
8th Floor, Lord Nelson Arcade
5675 Spring Garden Road
P.O. Box 3311
Halifax, Nova Scotia B3J 3J1
(902) 424-4014

ONTARIO
Workers' Compensation Board
2 Bloor Street East
Toronto, Ontario M4W 3C3
(416) 927-6968

PRINCE EDWARD ISLAND
Workers' Compensation Board
60 Belvedere Avenue
P.O. Box 757
Charlottetown, Prince Edward Island
C1A 7L7
(902) 368-5680

QUEBEC
Commission de la Sante et de la
Securite du travail
1199 de Bleury Street
P.O. Box 6056
Station A
Montreal, Quebec H3C 4E1
(514) 873-3503

SASKATCHEWAN
Workers' Compensation Board
1881 Scarth Street, #200
Regina, Saskatchewan S4P 2L8
(306) 787-4370

YUKON
Workers' Compensation Board
401 Strickland Street
Whitehorse, Yukon Y1A 4N8
(403) 667-5645

CANADA (Misc.)

Labour Canada
Federal Workers' Compensation
Service
Ottawa, Ontario K1A 0J2
(613) 997-2281

Merchant Seaman Compensation Board
Labour Canada
Ottawa, Ontario K1A 0J2

Appendix I

Chapter Review Questions

Study Guides

Chapter 1

1. Why was the workers' compensation system developed? When did it first come into being? Are all employees in the United States covered by Workers' Compensation Laws?
2. What are the major parts of the typical workers' compensation system? Describe their function in the system.
3. In what ways can an employer's immunity from civil suit be eliminated?
4. What state enacted the first and last workers' compensation laws?
5. What are the major types of claims?
6. There are six basic underlying objectives in a workers' compensation system. Name and describe them.
7. What types of cost factors are included within a business that has a workers' compensation claim?
8. What is the OSHA Act and what is its main purpose?

Chapter 2

1. What are the basic forms necessary for all employers to complete?
2. What types of questions might an employee be asking himself and why?
3. What is the purpose of keeping in touch with the employee while he is out on disability leave?
4. What are the basic steps an employer should take after an injury has occurred?
5. What are the three categories of workers' compensation jurisdictions? Describe each.

Chapter 3

1. What are the different types of injuries you need to be aware of? Give examples of the various types.
2. What is a repetitive or cumulative trauma? What is the importance of repetitive trauma to the employer?
3. What constitutes an injured worker?
4. Discuss the differences between cumulative or repetitive trauma and specific trauma.
5. What is meant by "arising out of employment" and "course of employment"?
6. What are the three different types of risks in employment?
7. What is the "going and coming rule"?
8. What is the purpose of a death claim?

Chapter 4

1. Give the names of some common repetitive motion injuries and the areas of the body affected.
2. What should you look for in the workplace that might cause a repetitive motion injury?
3. What are the basic stages of repetitive motion injuries?
4. What is carpal tunnel syndrome and what area of the body is affected?
5. What are some diseases that that are considered cumulative in nature and thus, compensable under workmen's compensation?
6. What are, statistically, the most common cumulative disorders?
7. What are the main causes of repetitive injury or cumulative trauma?

Chapter 5

1. Who are the different types of doctors typically included in a workers' compensation system?
2. What do you need to do when choosing a doctor or doctors for your business?
3. Why should you choose one type of doctor over another?
4. What is the most common repetitive injury?
5. Name a few different types of medical tests that your employee might have to undergo and describe what the tests are for.

Chapter 6

1. When does prevention begin in the workplace?
2. What things can you do to help prevent injuries in the workplace?
3. Why should employees be trained and then followed up on in regards to their job?
4. What are the different types of equipment used to help prevent injuries in the workplace?

Chapter 7

1. What is ergonomics and how is it important in the workplace?
2. Why is it necessary for an employee to have a proper chair?
3. What environmental hazards do employers need to look for?
4. Why is it important to have a good job design?
5. What are the basic steps involved in a good job design?
6. What types of ergonomic risk factors are critical in all businesses?

Chapter 8

1. Define disability and rehabilitation.
2. Define permanent total disability.
3. Define temporary total disability.
4. What is a work hardening program?
5. Describe what an employee might be feeling if he stays on disability for an extended period of time.
6. Know the basics of disability and their importance to managers or supervisors.
7. Who has control over the payment of benefits to the injured worker?
8. Why should you choose one insurance company over another?

Chapter 9

1. Who has control over the payment of benefits to the injured worker?
2. Why should you choose one insurance company over another?

Endnotes

Chapter 1. WHERE DID IT ALL BEGIN?

1. Arthur Larson. *The Law of Workmen's Compensation*, p.2-1. 1994, New York: Matthew Bender & Co., Inc. Reprinted with permission. All rights reserved.
2. Ibid., p. 2-5.
3. Ibid., p. 2-9.
4. Larson. pp. 2-22,23. Reprinted with permission.
5. Jeffrey Nackley. *A Primer on Workers' Compensation*, p. 85. 1989, Bureau of National Affairs, Inc., Washington, D.C. 20037. Reprinted with permission.
6. Larson. p. 2-22. Reprinted with permission.
7. Ibid, p. 2-23.
8. Nackley. p. 2. Reprinted with permission.
9. Ibid., p. 6-7.
10. U.S. Chamber of Commerce. *Analysis of Workers' Compensation Laws.* 1993, Washington, D.C..
11. CCH's *Workers' Compensation Manual for Managers and Supervisors*, p. 11. CCH Incorporated, 2700 Lake Cook Road, Riverwoods, Illinois 60015. 1992. Reprinted with permission.

Chapter 3. INJURIES AND THE INJURED WORKER

12. Nackley. p. 12. Reprinted with permission.
13. Ibid., pp. 14-16.
14. Ibid., pp. 11-17.

Chapter 4. TYPES OF INJURIES

15. Robert Bertolini. *Carpal Tunnel Syndrome, A Summary of the Occupational Health Concern.* 1990, Toronto: Canadian Centre for Occupational Health and Safety.
16. Bertolini. *Carpal Tunnel Syndrome, A Summary of the Occupational Health Concern.* 1990, Toronto: Canadian Centre for Occupational Health and Safety.
17. Nackley. pp. 25-27, 29-32. Reprinted with permission.

Chapter 5. MEDICAL CARE CHOICES

18. C. Taber. *Taber's Cyclopedic Medical Dictionary*. p. 31. 1981, F.A. Davis Company, Philadelphia.
19. American Chiropractic Association. *Chiropractic State of the Art*. pp. 32, 35. 1991-1992, Washington, D.C.
20. Ibid, p. 35.
21. American Chiropractic Association. pp. 40-41.

Chapter 6. PREVENTION

22. CCH's *Workers' Compensation Manual for Managers and Supervisors*. p. 23. CCH Incorporated, 2700 Lake Cook Road, Riverwoods, Illinois 60015. 1992, Reprinted with permission.
23. Ibid, p. 34.
24. Ilene Stones. *Ergonomics, A Basic Guide*. 1989, Toronto: Canadian Centre for Occupational Health and Safety.
25. Ibid.

Chapter 7. ERGONOMICS

26. Robert Arndt. *Workbook for Ergonomic Considerations in Office Design*. 1984, Alexandria, Virginia: National Office Products Association.
27. Ibid.
28. Ilene Stones. *Ergonomics: A Basic Guide*. 1989, Toronto: Canadian Centre for Occupational Health and Safety.
29. Arndt.
30. Ibid.
31. Andrew Drewczynski. *Working in a Standing Position*. Toronto: Canadian Centre for Occupational Health and Safety.
32. Ilene Stones. *Job Design: How it Contributes to Occupational Health and Safety*. 1989, Toronto:Canadian Centre for Occupational Health and Safety.
33. Ibid.
34. Ibid.
35. Ibid.

Chapter 8. DISABILITY AND REHABILITATION

36. Van Hemert MR. "Medical Factors and Disability Costs Associated with Long Term Low Back Disability Due to On-the-Job Injury," *The Digest of Chiropractic Economics*, July/August, 1993.
37. C. McGill. "Industrial Back Problems: A Control Program," *Journal of Occupational Medicine*. 1968, 10(4): 174-178.
38. *Nature of Injury by Occupation*. 1985, Toronto:Canadian Centre for Occupational Health and Safety.

Glossary

Glossary

Another aspect of workers' compensation is terminology. This is something that as an employer, it is not critical to thoroughly understand; however, having a general understanding of terms can be beneficial.

Accident — An event that happens unexpectedly that causes injury.

Aggravation — This term is used when an injured worker has experienced a new injury in the same area as a previous injury.

Amenable employer — Any employer who is subject to workers' compensation law and who is required to obtain workers' compensation coverage or, in those jurisdictions in which coverage is elective, who would be subject to penalties or loss of common-law defenses for failure to obtain coverage.

AOE/COE — Arising out of or occurring in the course of employment. This basically means that the employee must prove that the injury was sustained by reason of a condition or incident of the employment. An employee is considered as acting within the course of employment whenever he or she is carrying out some duty or right in connection with the employment.

Apportionment — When a previous injury has occurred and been rated by the labor board, the responsibility for payment is then divided between the previous injury and the new injury. This does not mean that the previous employer or insurance company will actually have an outlay of funds; rather, the second company is only liable for a portion of the payments.

Benefits — Any award paid under a claim, including compensation payable directly to claimants and payments for health-care providers, rehabilitation services, travel expenses, attorneys' fees, funeral payments, etc.

Burden of Proof — Degree of proof required to establish a claim. With a few exceptions, the burden of proof of each element of a compensable claim is on the claimant and is proof by a preponderance of the evidence. Elements of Fraud usually must be proved by clear and convincing evidence, and the burden is then on the party alleging fraud.

Claim — Any application for workers' compensation benefits filed by an injured or diseased worker or the dependents of a deceased worker.

Compensation — Those benefits available in workers' compensation systems payable directly to claimants in the form of wage loss, impairment in earnings, impairment in earning capacity, quasi-damages for loss of physical or mental functioning or loss of bodily parts, or, in death claims, survivors' benefits.

Cumulative Injury — (AOE) occurring as repetitive, mentally or physically traumatic or stressful activities extending over a period of time, which when combined causes a disability and/or need for medical treatment.

Death Claim — A claim for benefits usually filed by surviving dependents of a worker who allegedly was killed in the course of employment; also may be filed in most jurisdictions by non-dependent party who seeks reimbursement for bearing funeral or last-treatment costs of such worker.

Dependent — Those individuals defined by statutes as eligible for loss-of-support awards in death claims, usually including spouses, minor children, or other members of deceased worker's family who actually looked to decedent for support before the death, or before the injury or occupational disease that caused the death. Presumed dependents usually include spouses and minor children and, occasionally, parents. Permitted dependents sometimes include other members of decedent's family or household. Individuals not listed in statute as either presumed or permitted dependents generally are ineligible for loss-of-support benefits regardless of whether they were actually dependent upon the decedent for support before the death.

Disability — This is an inability to compete on the Open Labor Market, both present and future, after the point of maximum medical rehabilitation, due to functional restriction or pain that limits gainful work activity and work capacity.

Exacerbation — A worsening of an underlying disease or condition caused by work-related injury, accident, or occupational disease, usually thought of as a passing or one-time "flare-up" of the pre-existing problem, in contradistinction to aggravation.

Flare-up — This term is used when an injured worker has experienced an exacerbation in the same area as the current injury.

Fraud — An intentional falsehood upon which another party relies to his detriment, causing damages or change of position.

Going and coming rule — General rule of noncompensability for those injuries suffered by workers who have a fixed site of employment while going from home to regular place of employment or coming home from regular place of employment.

Impairment in earning capacity — Claimant's loss of ability to earn a living that has resulted from a work-related injury or occupational disease.

Independent contractor — An individual engaged by an entity to perform a job or a service but over whom the principal does not have the right to control the manner and means of performance. Independent contractors are not employees and, therefore, are generally not covered by workers' compensation.

Injury — (1) any physical or mental harm to an individual; (2) accident.

Jurisdiction — (1) the power of a government agency to act; (2) generic term for states, territories, commonwealths, the federal government, or other political entities.

Medical Impairment — Loss of physical or mental functioning which is determined strictly on the basis of medical evidence, including, but not limited to, clinical tests, objective findings, and result of physical examination.

Minimal exposure rule — Conditions for coverage of certain occupational diseases by which claimants must establish that they had been subject to injurious exposure for a specified period of time. Sometimes they must establish in-state exposure for specified period.

Non-complying employer — Amenable employer who has not complied with the workers' compensation law either by securing insurance coverage or by gaining approval to become self-insuring employer; usually liable for workers' compensation claims filed against its risk on dollar-for-dollar basis, and/or suable at common law without the common-law defenses of fellow servant rule, assumption of risk, and contributory negligence, and/or subject to fines and/or criminal penalties.

Objective Findings — These are findings that can be demonstrated or observed by the examiner such as: restriction of range of motion, muscle spasms, muscle atrophy, loss of strength, and X-ray findings.

Occupational Disease — An ailment that is contracted in the course of employment and is peculiar to an industrial trade or process in its causes, manifestations, frequency, or increased risk above that to which members of general public or of employment generally are exposed.

Onset-of-disability rule — Condition of coverage for certain occupational diseases by which claimants are required to establish that disability or death ensued within a specified period of time from date of last injurious exposure.

Permanent and Stationary — This is when an injured worker reaches a plateau in the treatment program where there will be no further benefit or relief from his residual symptomatology. At this point in time, the worker is evaluated and given a permanent disability rating. This term is only used when the physician expects some form of permanent disability. "Permanent and stationary" are always used together as though they were one word.

Permanent Partial Disability — When permanent disability is established after medical treatment has been given. This entitles the employee to receive compensation for the type and extent of his disability. There is a schedule in each state for compensation to the different areas of the body.

Permanent Total Disability — This is when an injured worker is unable to return to work. At this time he will receive benefits until the disability ceases. In some cases in some states, the employee will be given a lump settlement as opposed to receiving ongoing money.

Plaintiff — This is the legal term for the injured worker.

Rehabilitation — The process by which an injured worker is retrained once he is unable to return to his previous type of employment.

Scheduled disease — An occupational disease that is listed on statutory schedule as associated with particular industrial trade or process and, therefore, preemptively caused by employment or presumptively peculiar to industrial trade or process.

Second injury fund — Insurance fund, usually state-administered, under which costs or portions of costs of claims for aggravation of pre-existing conditions are paid rather than being charged to self-insured employer, employer's risk, or employer's insurance carrier.

Self-insuring employer — Complying employer who has obtained the right (or privilege) to pay directly and, to some extent, administer claims filed by its employees, i.e., without obtaining private-insurance or state fund coverage.

Self-procured treatment — Treatment that was initiated by the injured worker without the knowledge or approval of the employer or insurance company.

Specific Injury — (COE) injury occurring as a direct result of one incident or exposure which causes disability and/or need for medical treatment.

State fund — Insurance fund administered by state or public entity; may be competitive, i.e., private insurance is permitted in jurisdiction; or monopolistic, i.e., private insurance is not permitted in jurisdiction.

Subjective complaints — These are complaints experienced by the injured worker—for example, lower back pain, neck pain, arm, or leg pain. Also included in this definition would be symptoms such as numbness and burning.

Temporary Compensation — When an injured worker is unable to work due to a work injury, he is entitled to benefits which are equivalent to a percentage of his regular pay. This percentage varies from state to state, and there is always a maximum benefit. This gets paid to the employee until he is able to return to work.

Temporary Partial Disability — The period of time during which the injured worker may perform part of his work, or lighter duty, but cannot return to his regular work status.

Temporary Partial Payment — This entitles the employee to a percentage of the difference from what he is being paid at work and what he was being paid before the injury. For example, a full time employee working 40 hours, preinjury, can now only work 20 hours. He would be given a percentage of that difference in pay.

Temporary Total Disability — This is the period of time under which the employee is totally unable to return to his regular work duties.

Work-relatedness — General term for the casual connection that must be established by claimants between employment and disability. For occupational diseases, usually expressed by saying that disease must be "contracted in course of employment" and "peculiar to industrial trade or process"; for injuries, by saying that injury must be "in course of and arising out of employment".

Bibliography

Bibliography

Arndt, Robert. *Workbook for Ergonomic Considerations in Office Design*. Alexandria, Virginia: National Office Products Association, 1984.

Bertolini, Robert. *Carpal Tunnel Syndrome: A Summary of the Occupational Health Concern*. Toronto: Canadian Centre for Occupational Health and Safety, 1990.

Bertolini, Robert, and Andrew Drewczynski. *Repetitive Motion Injuries*. Toronto: Canadian Centre for Occupational Health and Safety, 1990.

Canadian Centre for Occupational Health and Safety, 250 Main Street, East Hamilton, Ontario L8N 1H6, (416) 572-2981.

CCH's *Workers' Compensation Manual for Managers and Supervisors*. CCH, Inc., 2700 Lake Cook Road, Riverwoods, Illinois 60015.

Chamber of Commerce of the United States from the 1995 Edition of *Analysis of Workers' Compensation Laws*. *Analysis of Workers' Compensation Laws* may be ordered by calling 1-800-638-6582.

Drewczynski, Andrew. *Working in a Standing Position*. Toronto: Canadian Centre for Occupational Health and Safety, 1990.

Health and Safety Products Catalog. North Coast Medical Inc., 187 Stauffer Boulevard, San Jose, CA 95125-1042.

Larson, Arthur. *The Law of Workmen's Compensation*. New York: Matthew Bender & Co., Inc. Copyright © 1994.

Nackley, Jeffrey V. *Primer on Workers' Compensation, Second Edition*. Copyright © 1989 by The Bureau of National Affairs, Inc., Washington, D.C. 20037.

Stones, Ilene. *Ergonomics: A Basic Guide*. Toronto: Canadian Centre for Occupational Health and Safety, 1989.

———. *Ergonomics for the Office*. Toronto: Canadian Centre for Occupational Health and Safety, 1989.

———. *Ergonomics for Workplaces with Visual Display Terminals*. Toronto: Canadian Centre for Occupational Health and Safety, 1989.

———. *Job Design: How it Contributes to Occupational Health and Safety*. Toronto: Canadian Centre for Occupational Health and Safety, 1989.

U.S. Department of Labor. *All About OSHA*. Washington, D.C., 1992.

Index